USED
15⁰⁰

GRADING TECHNIQUES FOR MODERN DESIGN

GRADING TECHNIQUES FOR MODERN DESIGN

Jeanne Price
Associate Professor: Fashion Design Department
Fashion Institute of Technology

Bernard Zamkoff
Chairman and Associate Professor: Fashion Design Department
Fashion Institute of Technology

FAIRCHILD PUBLICATIONS
NEW YORK

To J. Bicks and Howard

Copyright © 1974, by Fairchild Publications
Division Capital Cities Media, Inc.

Second Printing, 1978

Third Printing, 1983

All rights reserved. No part of this book may be reproduced in any
form without permission in writing from the publisher, except by
a reviewer who wishes to quote passages in connection with a
review written for inclusion in a magazine or newspaper.

SBN: 87005-102-4
Library of Congress Catalog Card Number: 73-8403

Printed in the United States of America

ACKNOWLEDGMENTS

The authors wish to give special recognition to John Rinaldi, friend and former student, for his early assistance, cooperation, and interest. And while the authors will have to take all the credit, or blame, for the drawings and diagrams in this book, they nevertheless are greatly indebted to Hugh Dobbins and Denise Serrante for their skilled technical work in preparing the illustrations for printing.

WHY THIS BOOK

The purpose of this book is to explain and demonstrate the skill of grading patterns by simple and reliable methods.

The information within is primarily for the student of design and for all those engaged in producing ready-to-wear clothes who realize the importance of knowing how to grade properly.

The methods illustrated in the following pages are fundamental and should be useful as a guide to grading patterns for all styles of misses and junior size garments.

This book will help you to a better understanding of the mechanics of grading, and thereby help you to produce garments with better fit for a greater variety of customers. That is, after all, the purpose, if not to say the magic, of grading.

The authors are confident that by following the methods in this book with reasonable practice and accuracy, a student can learn to apply the basic operations of grading to his or her other pattern-making skills.

CONVERSION TABLE (Inches to Centimeters)

			Inches					
Inches		1/8	1/4	3/8	1/2	5/8	3/4	7/8
		0.32	0.64	0.95	1.27	1.59	1.91	2.22
1	2.54	2.86	3.18	3.49	3.81	4.13	4.45	4.76
2	5.08	5.40	5.72	6.03	6.35	6.67	6.99	7.30
3	7.62	7.94	8.26	8.57	8.89	9.21	9.53	9.84
4	10.16	10.48	10.80	11.11	11.43	11.75	12.07	12.38
5	12.70	13.02	13.34	13.65	13.97	14.29	14.61	14.92
6	15.24	15.56	15.88	15.88	16.19	16.51	16.83	17.46
7	17.78	18.10	18.42	18.73	19.05	19.37	19.69	20.00
8	20.32	20.64	20.96	21.27	21.59	21.91	22.23	22.54
9	22.86	23.18	23.50	23.81	24.13	24.45	24.77	25.08
10	25.40	25.72	26.04	26.35	26.67	26.99	27.31	27.62
11	27.94	28.26	28.58	28.89	29.21	29.53	29.85	30.16
12	30.48	30.80	31.12	31.43	31.75	32.07	32.39	32.70
13	33.02	33.34	33.66	33.97	34.29	34.61	34.93	35.24
14	35.56	35.88	36.20	36.51	36.83	37.15	37.47	37.78
15	38.10	38.42	38.74	39.05	39.37	39.69	40.01	40.32
16	40.64	40.96	41.28	41.59	41.91	42.23	42.55	42.86
17	43.18	43.50	43.82	44.13	44.45	44.77	45.09	45.40
18	45.72	46.04	46.36	46.67	46.99	47.31	47.63	47.94
19	48.26	48.58	48.90	49.21	49.53	49.85	50.17	50.48
20	50.80	51.12	51.44	51.75	52.07	52.39	52.71	53.02
21	53.34	53.66	53.98	54.29	54.61	54.93	55.25	55.56
22	55.88	56.20	56.52	56.83	57.15	57.47	57.79	58.10
23	58.42	58.74	59.06	59.37	59.69	60.01	60.33	60.64
24	60.96	61.28	61.60	61.91	62.23	62.55	62.87	63.18
25	63.50	63.82	64.14	64.45	64.77	65.09	65.41	65.72
26	66.04	66.36	66.68	66.99	67.31	67.63	67.95	68.26
27	68.58	68.90	69.22	69.53	69.85	70.17	70.49	70.80
28	71.12	71.44	71.76	72.07	72.39	72.71	73.03	73.34
29	73.66	73.98	74.30	74.61	74.93	75.25	75.57	75.88
30	76.20	76.52	76.84	77.15	77.47	77.79	78.11	78.42
31	78.74	79.06	79.38	79.69	80.01	80.33	80.65	80.96
32	81.28	81.60	81.92	82.23	82.55	82.87	83.19	83.50
33	83.82	84.14	84.46	84.77	85.09	85.41	85.73	86.04
34	86.36	86.68	87.00	87.31	87.63	87.95	88.27	88.58
35	88.90	89.22	89.54	89.85	90.17	90.49	90.81	91.12
36	91.44	91.76	92.08	92.39	92.71	93.03	93.35	93.66
37	93.98	94.30	94.62	94.93	95.25	95.57	95.89	96.20
38	96.52	96.84	97.16	97.47	97.79	98.11	98.43	98.74
39	99.06	99.38	99.70	100.01	100.33	100.65	100.97	101.28
40	101.60	101.92	102.24	102.55	102.87	103.19	103.51	103.82
41	104.14	104.46	104.78	105.09	105.41	105.73	106.05	106.36
42	106.68	107.00	107.32	107.63	107.95	108.27	108.59	108.90
43	109.22	109.54	109.86	110.17	110.49	110.81	111.13	111.44
44	111.76	112.08	112.40	112.71	113.03	113.35	113.67	113.98
45	114.30	114.62	114.94	115.25	115.57	115.89	116.21	116.52
46	116.84	117.16	117.48	117.79	118.11	118.43	118.75	119.06
47	119.38	119.70	120.02	120.33	120.65	120.97	121.29	121.60
48	121.92	122.24	122.56	122.87	123.19	123.51	123.83	124.14
49	124.46	124.78	125.10	125.41	125.73	126.05	126.37	126.68
50	127.00	127.32	127.64	127.95	128.27	128.59	128.91	129.22
51	129.54	129.86	130.18	130.49	130.81	131.13	131.45	131.76
52	132.08	132.40	132.72	133.03	133.35	133.67	133.99	134.30
53	134.62	134.94	135.26	135.57	135.57	135.89	136.21	136.84
54	137.16	137.48	137.80	138.11	138.43	138.75	139.07	139.38
55	139.70	140.02	140.34	140.65	140.97	141.29	141.61	141.92
56	142.24	142.56	142.88	143.19	143.51	143.83	144.15	144.46
57	144.78	145.10	145.42	145.73	146.05	146.37	146.69	147.00
58	147.32	147.64	147.96	148.27	148.59	148.91	149.23	149.54
59	149.86	150.18	150.50	150.81	151.13	151.45	151.77	152.08
60	152.40	152.72	153.04	153.35	153.67	153.99	154.31	154.62
Inches		1/8	1/4	3/8	1/2	5/8	3/4	7/8

CONTENTS

PART 1
A STUDY OF GRADING

The Importance of Grading	1
Introductory Definitions	2
Body Landmarks	4
Types of Grade	5
Uneven Grade	7
Illustration of Figure Types	8
A Study of Figure Types	9
Using The Grading Chart	11
Skipping Sizes	13
Junior Bodice Chart	14
Misses Bodice Chart	15
Junior Sleeve Chart	16
Misses Sleeve Chart	17
Grading Machines	18
Blending Straight Lines and Corners	19
Blending Curves—neckline	20
Blending Curves—armhole	21
Back Bodice with shoulder dart	30

PART 2
GRADING BASIC DESIGNS

Guides to Good Grading	24
How The Body Grows—Front Bodice	25
Front Bodice	26
How The Body Grows—Back Bodice with shoulder dart	29
Back Bodice with shoulder dart	30
How The Body Grows—Back Bodice with neck dart	33
Back Bodice with neck dart	34
How The Body Grows—Skirt and Collars	37
Skirt and Collars	38
How The Body Grows—Torso	41
Torso	42
How The Body Grows—Slacks	45
Slacks	46
How The Body Grows—Set-in Sleeve	49
Set-in Sleeve	50

PART 3
GRADING INTERMEDIATE DESIGNS

How The Body Grows—Bodice and Yoke	55
Bodice and Yoke	56
How The Body Grows—Bodice and Midriff	59
Bodice and Midriff	60
How The Body Grows—Princess Bodice	63
Princess Bodice	64
How The Body Grows—Six Gore Skirt	67
Six Gore Skirt	68
How The Body Grows—Circle Skirt	71
Circle Skirt	72
Grading Skirt Variations	74
How The Body Grows—Shawl Collar	77
Shawl Collar	78

PART 4
GRADING ADVANCED DESIGNS

How The Body Grows—Kimono Sleeve	83
Kimono Sleeve	84
How The Body Grows—Kimono Raglan Sleeve	89
Kimono Raglan Sleeve	90
How The Body—Square Armhole Sleeve	95
Square Armhole	96
How The Body Grows—Gusset Sleeve	101
Gusset Sleeve	102
How The Body Grows—Set-in Raglan Sleeve	109
Set-in Raglan Sleeve	110
How The Body Grows—Princess Bodice and Sleeve in One	115
Princess Bodice and Sleeve in One	116
How The Body Grows—Dropped Shoulder Sleeve	121
Dropped Shoulder Sleeve	122
Additional Designs for Grading	126
Index	129

PART 1

A Study of Grading

THE IMPORTANCE OF GRADING

Every woman has had the experience of selecting from a rack of dresses marked in her size—only to find when trying them on that the fit varied from dress to dress. One of the reasons for this is that dresses are designed for a standard or model size and then graded up to larger sizes and down to smaller sizes, the results depending upon the skill of the grader. The grade is mechanically scaled up and down in proportion, but the human body unfortunately does not always grow according to standard measurements. Thus the proportion of the pattern will vary according to the experience, the accuracy, and the personal judgment of the grader. Each piece of the pattern for every new dress style has to be calculated, drawn, and cut out by hand while retaining the original lines that gave the garment its style. Errors can occur, and a dress that fit the model may end up not fitting the customer—who may then end up having a fit!

Therefore proper sizing is, first, very important to the ready-to-wear industry as it allows each manufacturer to produce a line of dresses capable of fitting a maximum number of women. Second, it is important to the clothing retailer in that it provides him with a wide range of sizes in each style and thus increases his potential market. Third, grading is important to the consumer, as it increases her chances of finding "her" dress in a size and style properly proportioned to her figure, and both she and the retailer need be involved in only a minimum of alterations.

Before beginning a study of grading, it is necessary to learn the definitions of some of the grading terms commonly used, as explained on the next pages.

INTRODUCTORY DEFINITIONS

GRADING | Grading is the process of proportionally increasing or decreasing a master pattern according to a prescribed set of body measurements. Each piece of pattern is shifted and traced step by step while at the same time the original style lines of the pattern are maintained. It is the skill, for instance, of changing a size 10 into a well fitting size 14, without losing the style proportions established by the designer.

MASTER PATTERN | The master pattern is the model size, or first pattern, from which duplicates are made and tested for accuracy. From these, production patterns are made with which the grader grades the remaining sizes up and down in proportion. The patternmaker devotes much time and effort to ensure that his master pattern will have perfect fit.

BODY MEASUREMENTS | The body measurements used in women's wear are not standardized as they are in children's and men's wear. Several attempts have been made at standardization but are not yet practical because of the greater variety of size and body types in women's ready-to-wear. The National Bureau of Standards has completed a project to provide scientific body measurements. Their analyses are based on actual body measurements, scientifically sampled and representing the predominant body types of the female population. As yet there is not a consensus on a fixed set of standard measurements that can be used as a general basis in grading, and thus there is a great deal of variation in the measurements now used. They are influenced and modified by the model formmakers, the Bureau of Standards, and in accordance with consumer complaints and suggestions.

GIRTH | The measurement of body circumferences, such as the bust, waist, and hip is referred to as girth. (It has the same derivation as the word "girdle"—that is, to encircle.)

SIZE RANGE | Under each standard size classification (see page 10), there is a group of at least six sizes, ranging from small to large, making a complete size designation. The misses sizes use even numbers while the junior size range uses odd numbers.

BODY LANDMARKS

LENGTH MEASUREMENTS

1-2. Center Front, neck to waistline
2-3. waistline to hip
3-4. hip to hemline
5-6. sideseam to waist
6-7. sideseam to hemline
5-8. underarm seam
9-10. overarm seam
11-12. Center Back, neck to waistline

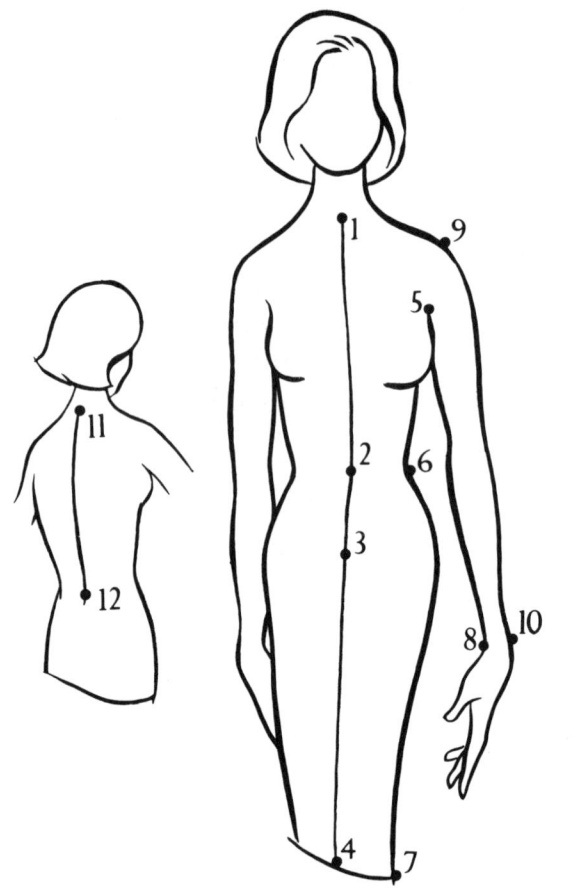

CIRCUMFERENCE AND WIDTH MEASUREMENTS

a. neck circumference
b. bust circumference
c. waist circumference
d. hip circumference
e. bicep circumference
f. elbow circumference
g. wrist circumference
h-i. cross shoulder width

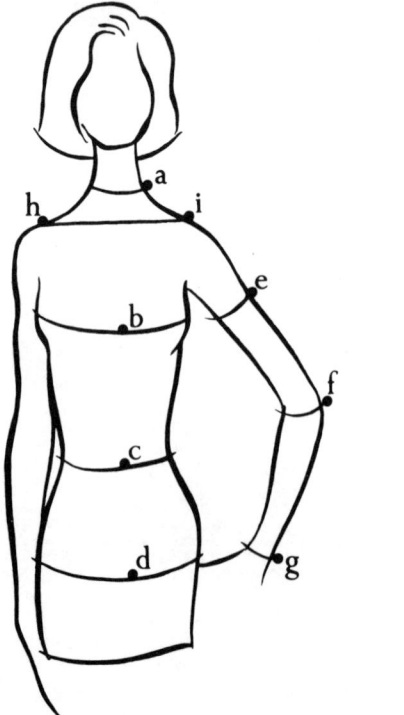

TYPES OF GRADE

GRADE | The increases or decreases of measurement between the sizes is referred to as the grade. The grade varies according to the type of measurement: circumference, length, or width.

CIRCUMFERENCE GRADE | The circumference grade indicates how much should be added to the front pattern in relation to the back pattern. To find the grade of a circumference measurement, the full measurement is taken of two circumference areas, and the difference between the measurements per size becomes the grade. Since we are working with the front pattern and the back pattern (each of which is one fourth of the whole circumference), this grade is divided into quarters, and just that quarter is added as a grade to the pattern.

EXAMPLE: If the Misses size 10 bust circumference measurement is 38½" and the size 12 is 40", the difference between the measurements (1½") becomes the grade. The 1½" grade means that on the front pattern, (one fourth of the whole garment) the width across the bust is increased by ⅜", and that the back pattern, at the same level, is increased by ⅜".

If the grade is 1", front and back are each increased ¼".

If the grade is 2", front and back are each increased ½".

LENGTH GRADE | The length grade indicates the measurement to be added to the length of the pattern. To find the grade of a length measurement, the difference between the length measurements per size becomes the grade. This is added on in proportion to the body's natural growth.

WIDTH GRADE | The width grade indicates the measurement to be added to a cross-body area, such as the distance from shoulder to shoulder across the chest. After finding the grade of a width area, divide the measurement in half, since only half of the total grade is needed. As a rule, the cross shoulder grade is always one-half of the bust grade.

EXAMPLE: If the bust grade is 1", the bust increase would be ¼". The cross shoulder grade is then one-half of that measurement, or ⅛". The cross shoulder increase itself is always distributed in two parts: a neck grade of $1/16$" and a shoulder grade equal to the remainder.

UNEVEN GRADE

All the grading directions in this book are for a 1½" grade on all areas of width. This means that the bust, waist, and hip all grade 1½" per size in the diagrams.

However, although the girth measurements of bust, waist, and hip are highly correlated with each other, the grade of these areas need not always be the same, and may indeed vary on all three of these girth measurements.

The following directions can be substituted at those times when an uneven grade is desired, such as in custom dressmaking and alterations.

BUST AND WAISTLINE

For an uneven grade between the bust and the waistline, the side seam would be graded in two movements of the pattern. (These directions are best studied after the even grade shown in the lesson has been practised and understood).

EXAMPLE: If you desire a bust grade of 1½" and a waistline grade of 1", Step 7 of the Front Bodice lesson would read: Shift the pattern down ⅛" from its previous position, and shift the pattern in until it is within ¼" of the length guideline. Trace the side seam-waistline intersection and blend the side seam.

SKIRT WAISTLINE AND HIP

The skirt waistline must be the same grade as the bodice waistline, although the hip grade may be larger.

EXAMPLE: If you desire a skirt waistline grade of 1½" and a hip grade of 2", Step 3 of the Skirt lesson, after establishing the 1½" waistline grade, would read: Trace the waistline-side seam intersection. Step 4 of the same lesson would then read: Shift the pattern down ¼" and out ⅛". Trace the side seam–hip intersection and blend the side seam to the waistline.

ILLUSTRATION OF FIGURE TYPES

JUNIOR PETITE	JUNIOR	MISSES	WOMEN	HALF-SIZES
5-7-9-11-13-15 size range	5-7-9-11-13-15 size range	6-8-10-12-14-16-18-20 size range	32-46 size range	12½-26½ size range
5' to 5'-4"	5'-4" to 5'-6"	5'-5" to 5'-8"	5'-5" to 5'-9"	under 5'-5"

A STUDY OF FIGURE TYPES

The garment industry produces millions of well fitting garments annually. This could not be done unless the existing size classifications were related to women of almost any figure type.

Every manufacturer begins with a master size pattern specifically made for the particular type of figure for which his styles were designed, as each figure type requires a specially developed pattern.

The opposite page illustrates the divisions that have been established into which these various figure types have been placed. Under each figure type division is their size and height range. The illustrations show the relative proportion of these body types by presenting them all on the same waist level.

The middle body type is the "average" Misses size range and has a well proportioned figure with a longer waist length than the slightly shorter Junior size woman to her left. The Junior body type has a youthful, high busted figure with a smaller waistline than the Misses. The more mature, well proportioned body type is found in the Women's size range, and has a fuller abdomen and lower bust than the Misses category.

The additional division of body types into half-sizes and Junior Petite sizes is a further step in the specialization of sizing in the ready-to-wear market. The half-sizes cater to the more mature, short-waisted woman with a shorter, heavier body type. The Junior Petite is for the woman with a small frame whose bust and hips are slightly smaller than the Junior, and whose waist length is slightly shorter.

The lessons in this book concentrate on the popular Misses size range, in the 1½" grade. However, all the lessons can be applied to the other size ranges, as each lesson makes note of those steps which need adjusting for the 1" grade of the Junior Petite and Junior sizes, and the 2" grade of the Womens and half-size range. As a guide for the grading of Junior size patterns, charts for the Junior size range are included in this book.

USING THE GRADING CHART

A grading system begins with a grading chart. Prepared with great care, the grading chart reflects the size ranges that the manufacturer regards as his market. The purpose of the grading chart is to indicate to the grader the differences in circumference, length, and width measurements of the various parts of the pattern as they increase or decrease in size.

The following grading charts show the measurement differences, or grade, between the various sizes. The data for these charts have been compiled from a study which lists the full measurements of all the areas of the body for all sizes.

The charts are based on the following idea: a woman's bust increases in circumference either 1", 1½", or 2" as she grows larger. Generally, the smaller sizes grow 1", the middle range 1½", and the larger sizes 2". In this book we have assigned the 1" increase, or grade as it is called, to sizes 6 through 10 and 5 through 9, the 1½" grade to sizes 10 through 18 and 9 through 15, and the 2" grade to sizes 18 to 20. This division of the size ranges is not necessarily the one used by all manufacturers, but it is one that we have found to be both reasonable and accurate.

The bust grade and the areas dependent upon it (the cross shoulder, waistline, and hip) are the only variable measurements from one size to the next. All other bodice and skirt grades are standard per size. The variable grade on a sleeve is the bicep width, which is also dependent upon the bust grade. The charts take into account all of these variable areas, as well as those areas which grade a standard amount per size.

To further aid the student, the numbered columns on the bodice charts correspond to the numbered boxes on the grading lesson of the Front Bodice on pages 26 and 27.

TO GRADE

To use the following grading charts, first find the size of your production pattern in the column marked "GRADE". Next choose the size to which you want to increase your pattern under the column marked "DOWNGRADE". Read the chart from left to right following the line of measurements which ends with the size you have chosen. The 1", 1½", or 2" grade for your size is indicated under the cross shoulder column and cross bust column of the bodice chart, and under the bicep, wrist and waistline columns of the sleeve and skirt chart.

EXAMPLE: To increase a size 10 pattern to a size 12, locate size 10 in the "GRADE" column. Follow the line of bodice measurements from the left to size 12 on the right in the "DOWNGRADE" column, and use those measurements in your grading. As a grade from a size 10 to a size 12 is a 1½" grade, you would use the measurements given under that heading in the cross shoulder and cross bust columns.

TO DOWNGRADE

To downgrade, first find the size of your production pattern in the column marked "DOWNGRADE". Next, choose the size to which you want to decrease your pattern under the column marked "GRADE". Read the chart from right to left following the line of measurements which ends with the size you have chosen. All the directions in this book are for increasing your pattern size, so to downgrade a pattern the directions would have to be reversed.

EXAMPLE: To decrease a size 12 pattern to a size 10, locate the size 12 in the "DOWNGRADE" column that is lined up with the size 10 in the "GRADE" column. Decrease your pattern by reversing the directions, using the measurements on that line.

SKIPPING SIZES

For beginners, greater accuracy and better proportion can be achieved if the pattern is graded from size to size, tracing each size as you go until you reach the size desired. It is important to always grade accurately and in proportion since each grade is dependent upon the one before it. With practice, you will find that you can easily skip a size or more, going directly from your master size to the size desired, and the charts are arranged so that the pattern can be graded in that manner. For instance, a size 6 pattern can be graded directly up to a size 10 by using the measurements on the line connecting size 6 on the left with size 10 on the right. However, it is not good practice to skip more than a couple of sizes in most instances, as proportion could get lost.

BODICE GRADING CHART - Junior Sizes

Grade	Shoulder Level Grade	Neck Grade	Cross Shoulder Grade 1"	Cross Shoulder Grade 1½"	Armhole Grade	Cross Bust Grade 1"	Cross Bust Grade 1½"	Side Seam Grade	Apex Width Grade	Dart Length Grade	Downgrade
5	1/8	1/16	1/8		APPROX. 3/16	1/4		1/8	1/8	1/8	7
5	1/4	1/8	1/4		3/8	1/2		1/4	1/4	1/4	9
5	3/8	3/16		7/16	9/16		7/8	3/8	3/8	3/8	11
5	1/2	1/4		5/8	3/4		1 1/4	1/2	1/2	1/2	13
5	5/8	5/16		13/16	15/16		1 5/8	5/8	5/8	5/8	15
7	1/8	1/16	1/8		3/16	1/4		1/8	1/8	1/8	9
7	1/4	1/8		5/16	3/8		5/8	1/4	1/4	1/4	11
7	3/8	3/16		1/2	9/16		1	3/8	3/8	3/8	13
7	1/2	1/4		11/16	3/4		1 3/8	1/2	1/2	1/2	15
9	1/8	1/16		3/16	3/16		3/8	1/8	1/8	1/8	11
9	1/4	1/8		3/8	3/8		3/4	1/4	1/4	1/4	13
9	3/8	3/16		9/16	9/16		1 1/8	3/8	3/8	3/8	15
11	1/8	1/16		3/16	3/16		3/8	1/8	1/8	1/8	13
11	1/4	1/8		3/8	3/8		3/4	1/4	1/4	1/4	15
13	1/8	1/16		3/16	3/16		3/8	1/8	1/8	1/8	15

BODICE GRADING CHART - Misses Sizes

GRADE	2 SHOULDER LEVEL GRADE	3 NECK GRADE	4 CROSS SHOULDER GRADE 1"	4 CROSS SHOULDER GRADE 1½"	4 CROSS SHOULDER GRADE 2"	5 ARMHOLE GRADE	6 CROSS BUST GRADE 1"	6 CROSS BUST GRADE 1½"	6 CROSS BUST GRADE 2"	7 SIDE SEAM GRADE	8 APEX WIDTH GRADE	9 DART LENGTH GRADE	DOWNGRADE
6	1/8	1/16	1/8			APPROX. 3/16	1/4			1/8	1/8	1/8	8
6	1/4	1/8	1/4			3/8	1/2			1/4	1/4	1/4	10
6	3/8	3/16		7/16		9/16		7/8		3/8	3/8	3/8	12
6	1/2	1/4		5/8		3/4		1 1/4		1/2	1/2	1/2	14
6	5/8	5/16		13/16		15/16		1 5/8		5/8	5/8	5/8	16
6	3/4	3/8		1		1 1/8		2		3/4	3/4	3/4	18
6	7/8	7/16			1 1/4	1 5/16			2 1/2	7/8	7/8	7/8	20
8	1/8	1/16	1/8			3/16	1/4			1/8	1/8	1/8	10
8	1/4	1/8		5/16		3/8		5/8		1/4	1/4	1/4	12
8	3/8	3/16		1/2		9/16		1		3/8	3/8	3/8	14
8	1/2	1/4		11/16		3/4		1 3/8		1/2	1/2	1/2	16
8	5/8	5/16		7/8		15/16		1 3/4		5/8	5/8	5/8	18
8	3/4	3/8			1 1/8	1 1/8			2 1/4	3/4	3/4	3/4	20
10	1/8	1/16		3/16		3/16		3/8		1/8	1/8	1/8	12
10	1/4	1/8		3/8		3/8		3/4		1/4	1/4	1/4	14
10	3/8	3/16		9/16		9/16		1 1/8		3/8	3/8	3/8	16
10	1/2	1/4		3/4		3/4		1 1/2		1/2	1/2	1/2	18
10	5/8	5/16			1	15/16			2	5/8	5/8	5/8	20
12	1/8	1/16		3/16		3/16		3/8		1/8	1/8	1/8	14
12	1/4	1/8		3/8		3/8		3/4		1/4	1/4	1/4	16
12	3/8	3/16		9/16		9/16		1 1/8		3/8	3/8	3/8	18
12	1/2	1/4			13/16	3/4			1 5/8	1/2	1/2	1/2	20
14	1/8	1/16		3/16		3/16		3 1/8		1/8	1/8	1/8	16
14	1/4	1/8		3/8		3/8		3/4		1/4	1/4	1/4	18
14	3/8	3/16			5/8	9/16		1 3/8		3/8	3/8	3/8	20
16	1/8	1/16		3/16		3/16		3/8		1/8	1/8	1/8	18
16	1/4	1/8			7/16	3/8			7/8	1/4	1/4	1/4	20
18	1/8	1/16			1/4	3/16			1/2	1/8	1/8	1/8	20

SLEEVE & SKIRT GRADING CHART
Junior Sizes

GRADE	BICEP GRADE AND ELBOW GRADE		CAP HEIGHT GRADE	UNDERARM SEAM BICEP TO WRIST	ELBOW DART LENGTH GRADE	FITTED WRIST		STRAIGHT WRIST		SKIRT WAISTLINE		DOWNGRADE
	1"	1½"				1"	1½"	1"	1½"	1"	1½"	
5	3/8		1/8	1/4	3/32	3/16		3/8		1/4		7
	3/4		1/4	1/2	3/16	3/8		3/4		1/2		9
		1¼	3/8	3/4	5/16		5/8		1¼		7/8	11
		1¾	1/2	1	7/16		7/8		1¾		1¼	13
		2¼	5/8	1¼	9/16		1⅛		2¼		1⅝	15
7	3/8		1/8	1/4	3/32	3/16		3/8		1/4		9
		7/8	1/4	1/2	7/32		7/16		7/8		5/8	11
		1⅜	3/8	3/4	11/32		11/16		1⅜		1	13
		1⅞	1/2	1	15/32		15/16		1⅞		1⅜	15
9		1/2	1/8	1/4	1/8		1/4		1/2		3/8	11
		1	1/4	1/2	1/4		1/2		1		3/4	13
		1½	3/8	3/4	3/8		3/4		1½		1⅛	15
11		1/2	1/8	1/4	1/8		1/4		1/2		3/8	13
		1	1/4	1/2	1/4		1/2		1		3/4	15
13		1/2	1/8	1/4	1/8		1/4		1/2		3/8	15

SLEEVE & SKIRT GRADING CHART
Misses Sizes

GRADE	1 BICEP GRADE AND ELBOW GRADE			2 CAP HEIGHT GRADE	3 UNDER-ARM SEAM BICEP TO WRIST	4 ELBOW DART LENGTH GRADE	5 FITTED WRIST			6 STRAIGHT WRIST			7 SKIRT WAISTLINE			DOWN GRADE
	1"	1½"	2"				1"	1½"	2"	1"	1½"	2"	1"	1½"	2"	
6	3/8			1/8	1/4	3/32	3/16			3/8			1/4			8
	3/4			1/4	1/2	3/16	3/8			3/4			1/2			10
		1¼		3/8	3/4	5/16		5/8			1¼			7/8		12
		1¾		1/2	1	7/16		7/8			1¾			1¼		14
		2¼		5/8	1¼	9/16		1⅛			2¼			1⅝		16
		2¾		3/4	1½	11/16		1⅜			2¾			2		18
			3⅜	7/8	1¾	27/32			1 11/16			3⅜			2½	20
8	3/8			1/8	1/4	3/32	3/16			3/8			1/4			10
		7/8		1/4	1/2	7/32		7/16			7/8			5/8		12
		1⅜		3/8	3/4	11/32		11/16			1⅜			1		14
		1⅞		1/2	1	15/32		15/16			1⅞			1⅜		16
		2⅜		5/8	1¼	19/32		1 3/16			2⅜			1¾		18
			3	3/4	1¾	3/4		1½				3			2¼	20
10		1/2		1/8	1/4	1/8		1/4			1/2			3/8		12
		1		1/4	1/2	1/4		1/2			1			3/4		14
		1½		3/8	3/4	3/8		3/4			1½			1⅛		16
		2		1/2	1	1/2		1/2			2			1½		18
			2⅝	5/8	1¼	21/32		1 5/16				2⅝			2	20
12		1/2		1/8	1/4	1/8		1/4			1/2			3/8		14
		1		1/4	1/2	1/4		1/2			1			3/4		16
		1½		3/8	3/4	3/8		3/4			1½			1⅛		18
			2⅛	1/2	1	17/32		1 1/16				2⅛			1⅝	20
14		1/2		1/8	1/4	1/8		1/4			1/2			3/8		16
		1		1/4	1/2	1/4		1/2			1			3/4		18
		1⅝		3/8	3/4	13/32		13/16			1⅝			1⅜		20
16		1/2		1/8	1/4	1/8		1/4			1/2		3/8			18
		1⅛		1/4	1/2	9/32		9/16			1⅛			7/8		20
18				5/8	1/8	1/4	5/32		5/16			5/8			1/2	20

GRADING MACHINES

A machine is not a substitute for grading knowledge, but if it is used properly it will enable the grader to have greater speed and accuracy. There are various hand machines on the market, but in general they all have the same format and are simple to use.

A grading machine can grade one size at a time without removing the original pattern from the machine. Greater speed can be achieved by grading two parts of the pattern at the same time. For example, the back and front bodice can be graded in one movement as shown in the diagram. Knobs control a set of scales that represent the girth and length measurements and mechanically shift the pattern in horizontal and vertical movements. To obtain the larger sizes, the measurements are multiplied. When all the sizes are graded, the pattern must be removed from the machine to blend or join points on curves and angles.

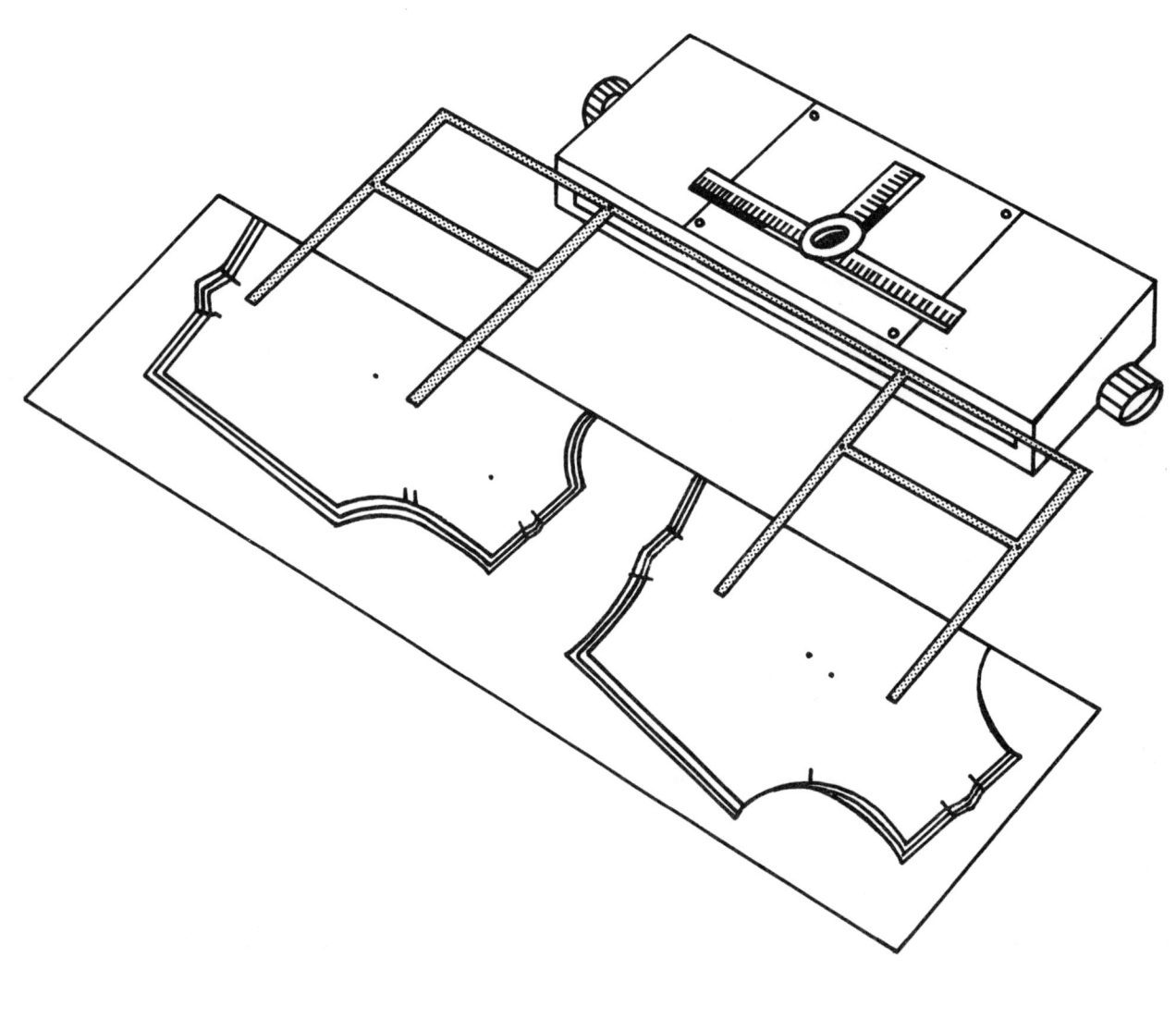

BLENDING

In grading the pattern, disconnected lines may occur in certain areas. These disconnected lines must be blended into one free-flowing continuous line.

BLENDING STRAIGHT LINES

1. When disconnected lines occur in a straight area of the pattern, choose a central line between them and blend into one continuous line.
EXAMPLE: Sleeve underarm seam, shoulder seam.

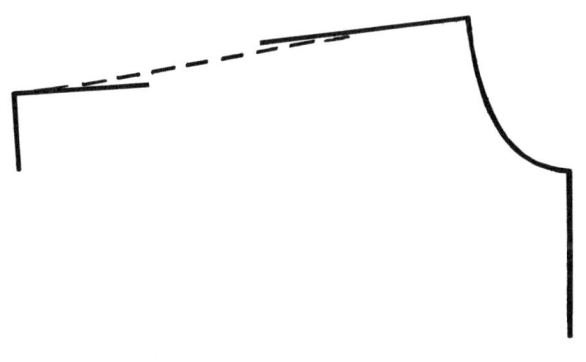

BLENDING CORNERS

2. When blending an unconnected corner, extend a line across to form an intersection at a right angle to the other line.
EXAMPLE: Waistline-Center Front and Center-Back intersections.

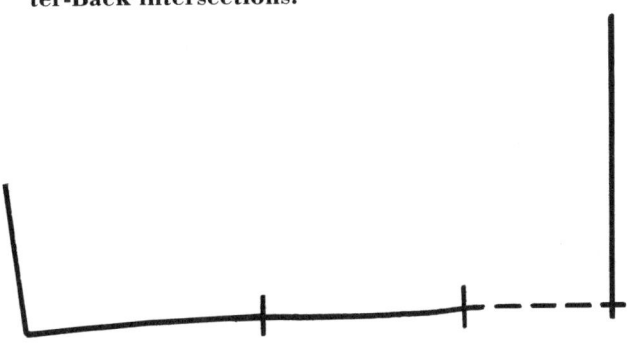

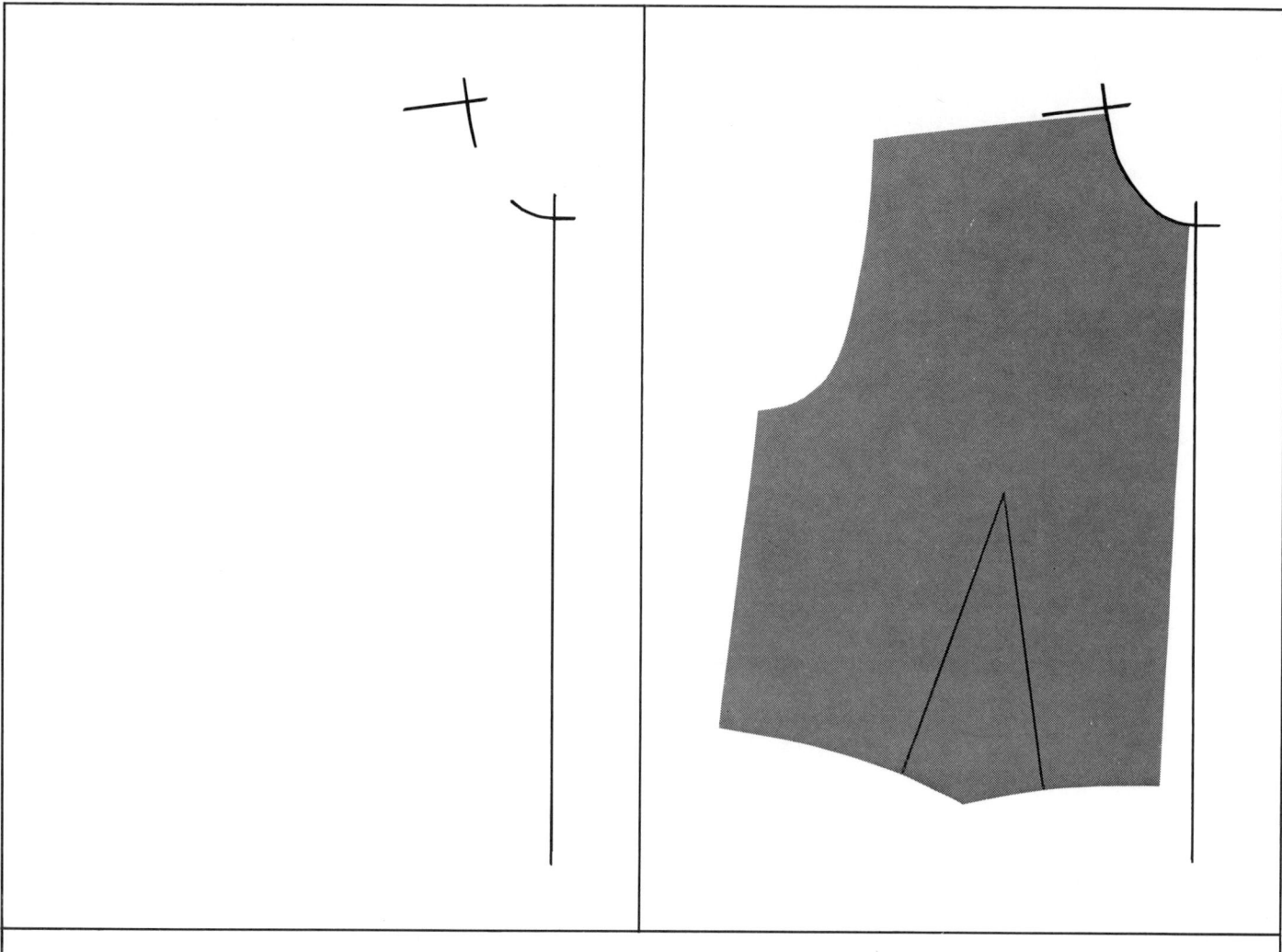

BLENDING CURVES

When blending a curved line, use the exact pattern section as a guide. The diagrams illustrate blending the neckline and armhole curves, and may also be used as a guide for blending all curved lines such as sleeve caps, crotch seams, and the outer edges of collars.

BLENDING THE NECKLINE

DIAGRAM 1. illustrates the tracing of the developing front bodice pattern before the neckline has been blended.

DIAGRAM 2. shows the blending of the neckline. Place the original front bodice pattern in such a position that its neck edge makes a smooth continuation with the traced portions of diagram 1. Then trace the pattern's neck edge, connecting it to the traced portions. Lift off the pattern. The blended curve from Center Front to the shoulder is now the graded neckline of the developing front bodice pattern. The back bodice neckline is blended in a similar manner to the front.

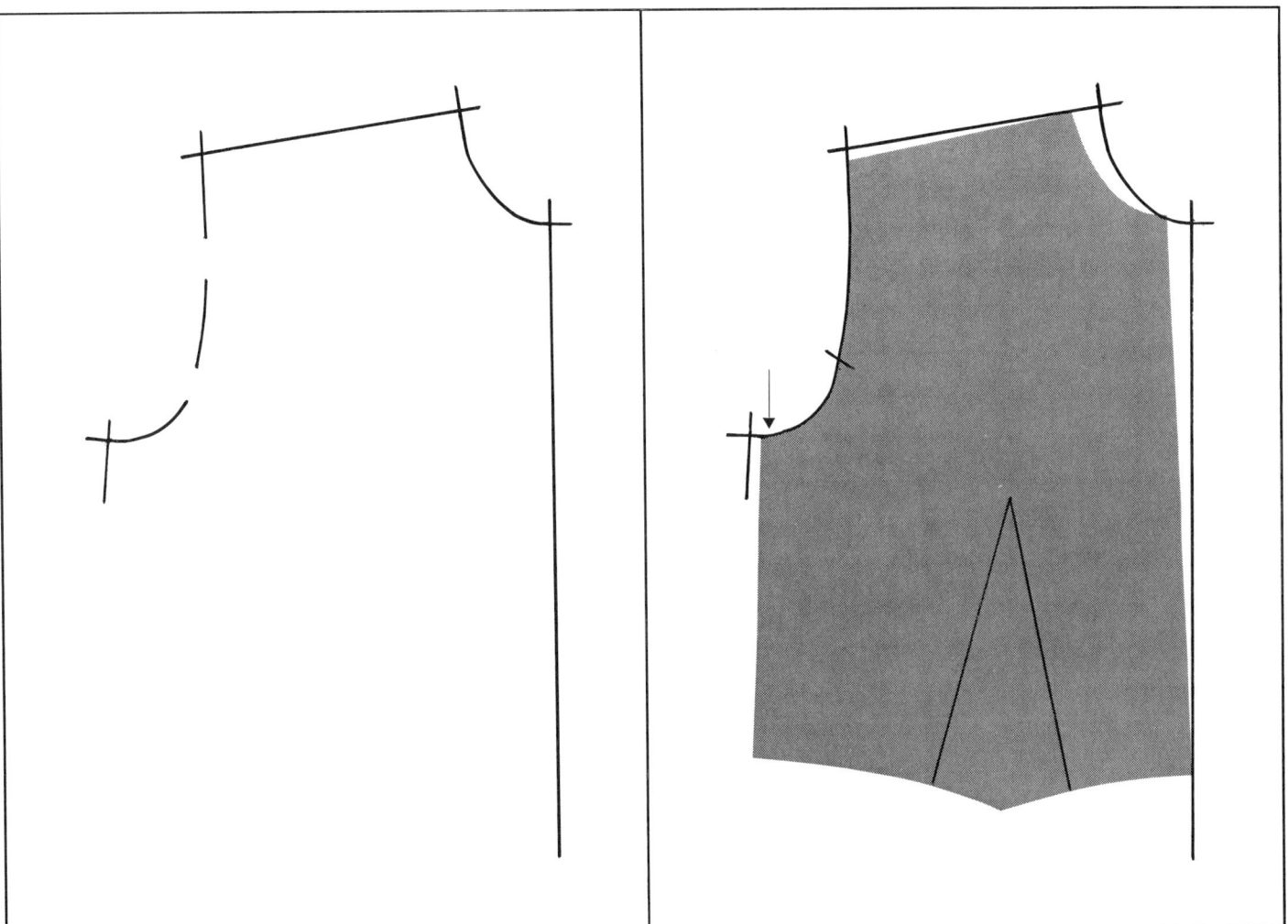

BLENDING THE ARMHOLE

DIAGRAM 1. illustrates the tracing of the developing front bodice pattern before the armhole has been blended.

DIAGRAM 2. shows the blending of the armhole. Overlay the master front bodice pattern so that the part of the curve containing the notch is spaced a distance away from the developing pattern an amount equal to the grade at that part of the curve, as shown by arrow in diagram II. For example: on a front bodice 1½" grade, the master pattern would be placed 3/16" away from the side seam of the developing pattern. Then trace the pattern's armhole edge, connecting it to the traced portions, and mark the armhole notch. Lift off the pattern. The blended curve from the shoulder to the side seam is now the graded armhole of the developing front bodice pattern. The back bodice armhole is blended in the same manner as the front.

PART 2

Grading Basic Designs

BODICE

SKIRT

COLLARS

TORSO

SLACKS

SET-IN SLEEVE

GUIDES TO GOOD GRADING

Each grading lesson in this book is preceded by a split diagram designed to help the reader visualize how the body grows. All the diagrams illustrate the 1½″ grade, and are closely correlated to the shifts of the pattern in the step-by-step directions which follow each diagram.

The reader is urged to study carefully the split diagram before beginning a lesson, in order to become thoroughly familiar with how the body grows in each of its areas. The notes accompanying each diagram further clarify the particular points of importance in each individual lesson.

Before beginning actual grading, all pattern parts must be assembled and inspected for accuracy. The successful proportions of the graded sizes depends upon the perfection of the original pattern.

Always shift the pattern parallel to the guidelines indicated in each lesson.

Use great care and accuracy in the shifting of the pattern, whether grading by hand or machine, following the measurements given in the lesson.

Steps marked with an asterisk have a special note at the top of the lesson which will help you apply that lesson to the 1″ or 2″ grades.

HOW THE BODY GROWS

FRONT BODICE
1½" Grade

1. The cross shoulder grade of the bodice is always one-half of the cross bust grade.

2. The neck increase of the bodice appears to be ³⁄₁₆", but it actually will be ⅛" once the neckline is blended into a smooth curve.

3. The ⅛" increase at the neckline becomes ¹⁄₁₆" at the armhole due to maintaining the shoulder level while grading, as explained in the directions.

4. The ⅛" increase at the shoulder is evenly distributed at the waistline as ¹⁄₁₆" on either side of the dart. The ¹⁄₁₆" closest to Center Front joins the neck increase of ¹⁄₁₆" to become a ⅛" increase at the waistline.

5. An armhole increases in depth a standard ³⁄₁₆" per size, and increases in width a variable amount equal to one half of the cross bust grade.

6. The standard apex width grade of ⅛" per size may also be graded as a variable, equal to one-half of the cross bust grade.

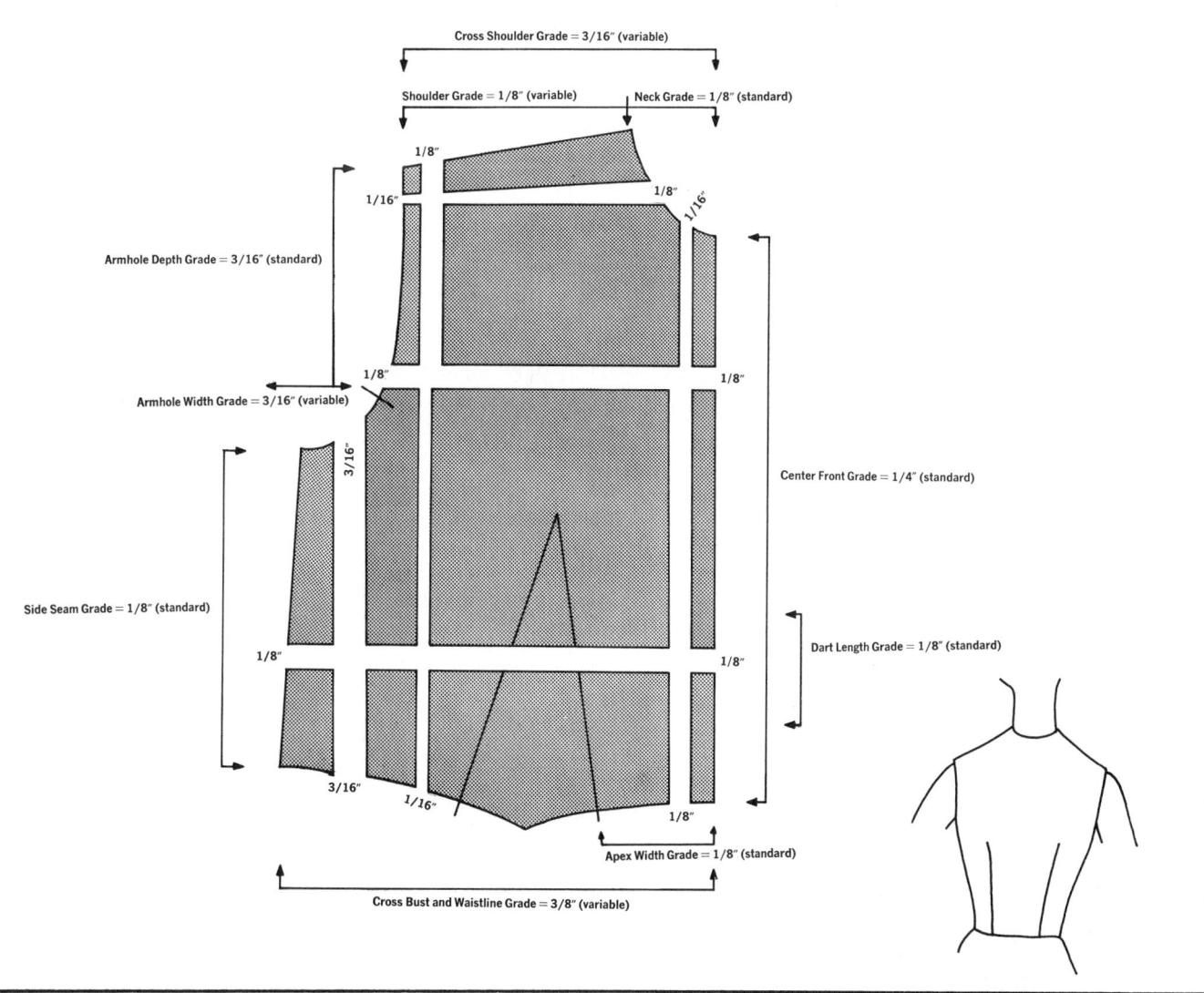

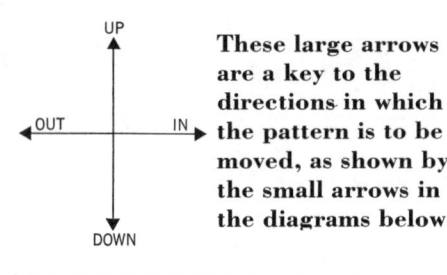

These large arrows are a key to the directions in which the pattern is to be moved, as shown by the small arrows in the diagrams below.

DIRECTIONS FOR A FRONT BODICE 1½″ GRADE

1 CENTER FRONT

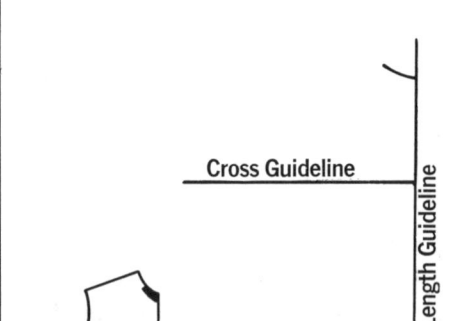

Draw a cross control line on the pattern by squaring a line from Center Front to the underarm—side seam intersection. On paper, draw a length guideline, and square a cross guideline from it. Place Center Front of the pattern on the length guideline, with the cross control line of the pattern on the cross guideline. Trace the neckline—Center Front intersection. (The length guideline will be the Center Front of the graded pattern).

2 SHOULDER LEVEL GRADE

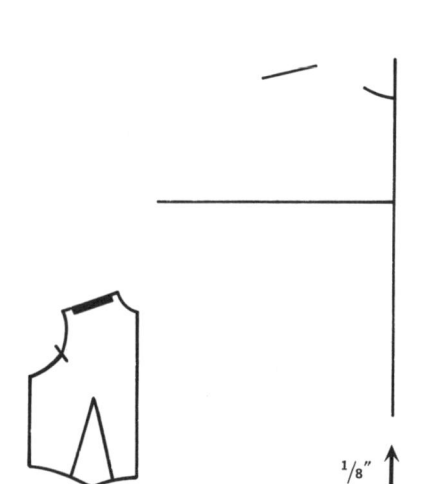

⅛″ ↑

4 CROSS SHOULDER GRADE*

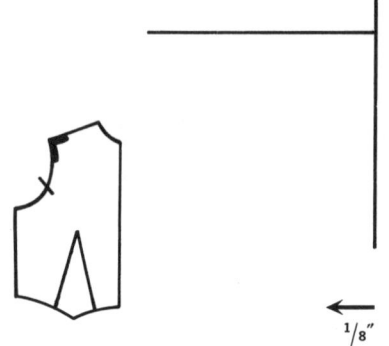

← ⅛″

Shift the pattern out ⅛″ from its previous position, maintaining shoulder level. The pattern should now be a total of ³⁄₁₆″ from the length guideline, and the cross control line of the pattern should now be approximately ¹⁄₁₆″ above the cross guideline on the paper. Trace the remainder of the shoulder, and the armhole intersection. Blend the shoulder seam as shown on page 19.

NOTE: The maintaining of shoulder level in the cross shoulder step of any bodice is used only for an 1½″ grade. Therefore, when doing a cross shoulder step on a 1″ or 2″ grade, the cross control line of pattern must be placed ¹⁄₁₆″ above the cross guideline on the paper, disregarding shoulder level, before tracing the shoulder-armhole intersection.

5 ARMHOLE GRADE

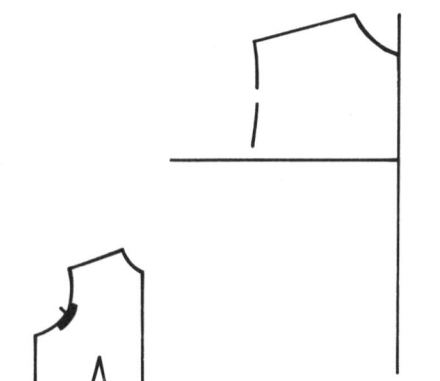

approx.
↓ ³⁄₁₆″

7 SIDE SEAM GRADE

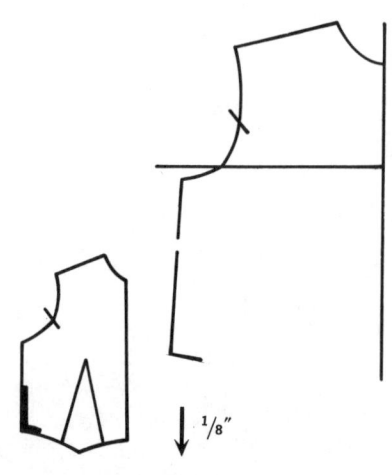

↓ ⅛″

Shift the pattern down ⅛″ from its previous position. The cross control line of the pattern should now be a total of ¼″ below the cross guideline on the paper. Trace the side seam to and around the waistline intersection.

8 APEX WIDTH GRADE

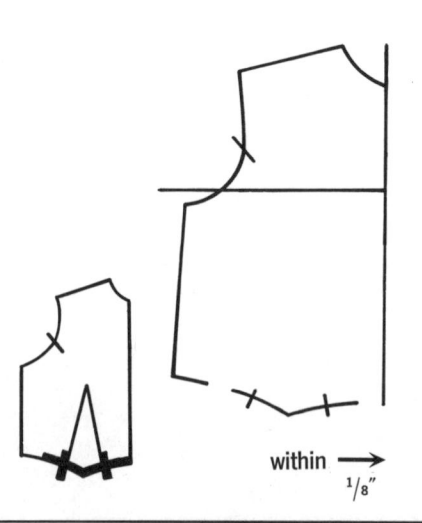

within ⅛″ →

1. In all steps except 4* and 6*, the measurements used are standard from one size to the next, no matter if it be a 1″, 1½″, or 2″ grade. Refer to the bodice grading chart for the measurements for steps 4 and 6 *when doing other than the 1½″ grade shown below*. For step 4, subtract column 3 from column 4 before applying the measurement. For step 6, divide the measurement in column 19 in half before using.

2. The smaller diagram in the lower corner of each box depicts the pattern being graded, with the section to be traced in that step marked in heavy lines.

Shift the pattern up on the length guideline, until the cross control line of the pattern is ⅛″ above the cross guideline on the paper. Trace the middle of the shoulder seam in order to establish the shoulder level.

3 NECK GRADE

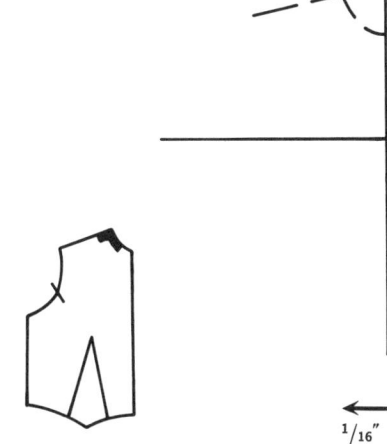

← 1/16″

Shift the pattern out 1/16″ from the length guideline. Before tracing, adjust the pattern so that the shoulder seam is once again directly on the shoulder level traced in step 2. This adjustment is known as "maintaining shoulder level." Now trace the shoulder-neckline intersection. Blend the neckline as shown on page 20.

Shift the pattern down so that the cross control line of the pattern is ⅛″ below the cross guideline on the paper. The total downward shift of the pattern from its previous position will be approximately 3/16″. Trace the middle of the armhole.

6 CROSS BUST GRADE*

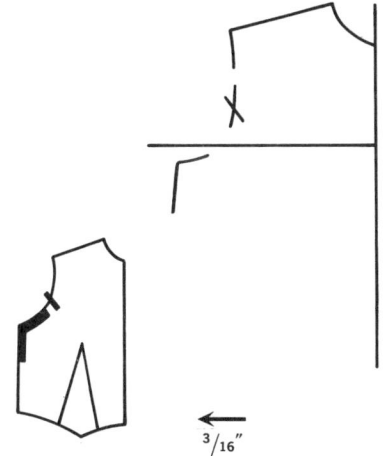

← 3/16″

Shift the pattern out 3/16″ from its previous position. The pattern should now be a total of ⅜″ from the length guideline. Trace the lower armhole and side seam intersection. Blend the armhole as shown on page 21. Mark the armhole notch.

Shift the pattern in until it is within ⅛″ of the length guideline. Trace the waistline, and mark the dart notches. Blend the waistline across to the length guideline, thus establishing the new Center Front—waistline intersection. (See page 19).

9 DART LENGTH GRADE

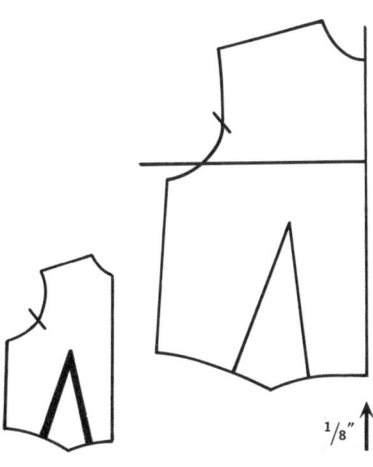

↑ ⅛″

With the pattern still within ⅛″ of the length guideline, shift the pattern up ⅛″ from its previous position and mark the dart apex. Lift the pattern off the paper and draw in the dart lines. The front bodice 1½″ grade is now complete.

HOW THE BODY GROWS

BACK BODICE WITH SHOULDER DART

1½" Grade

1. The back bodice grade must be consistent with the front bodice grade.
2. The shoulder dart should remain centered; therefore the ⅛" shoulder in-increase for the 1½" grade should be divided in half, with ¹⁄₁₆" being placed on either side of the dart.

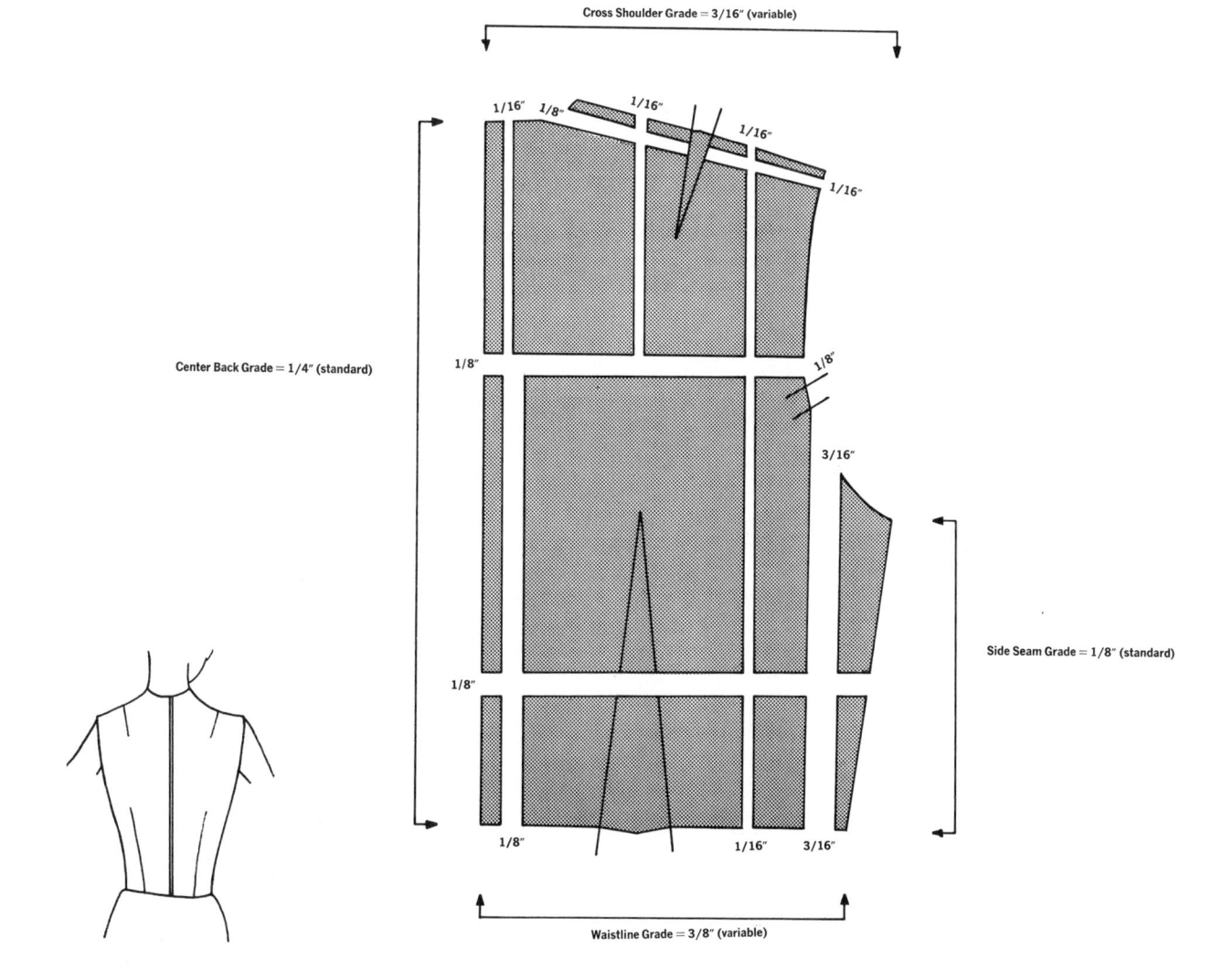

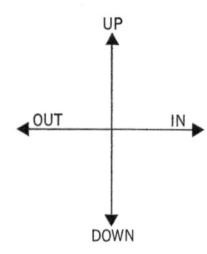

These large arrows are a key to the directions in which the pattern is to be moved, as shown by the small arrows in the diagrams below.

DIRECTIONS FOR A BACK BODICE 1½ INCH GRADE (SHOULDER DART)

1 CENTER BACK

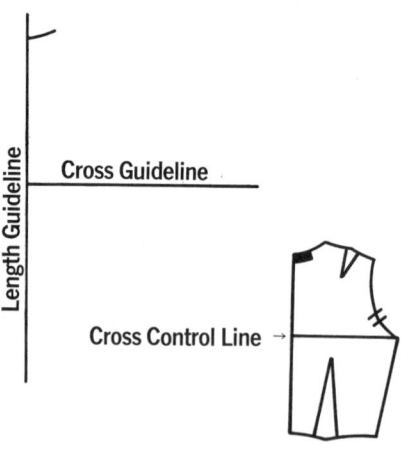

Draw a cross control line on the pattern by squaring a line from Center Back to the underarm-side seam intersection. On paper, draw a length guideline, and square a cross guideline from it. Place Center Back of the pattern on the length guideline, with the cross control line of the pattern on the cross guideline. Trace the neckline-Center Back intersection. (The length guideline will be the Center Back of the graded pattern).

2 SHOULDER LEVEL GRADE

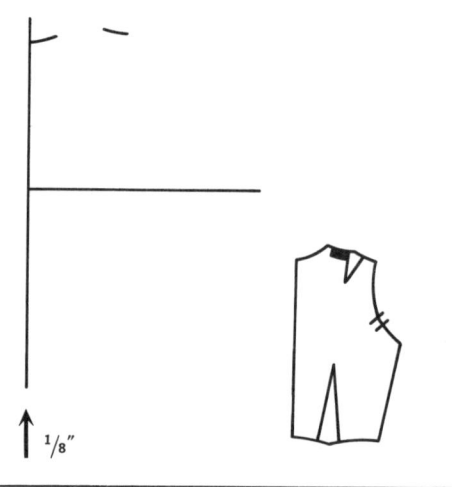

4 CROSS SHOULDER GRADE*

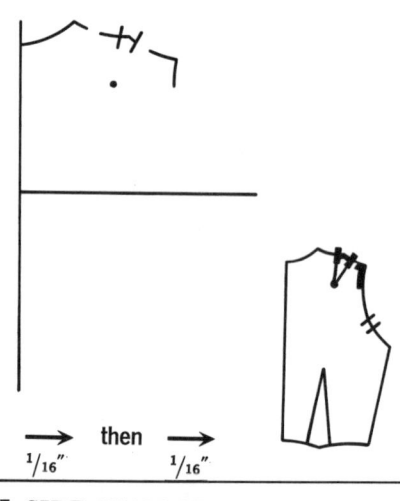

Shift the pattern in $1/16"$ from its previous position, maintaining shoulder level. The pattern should now be a total of $1/8"$ from the length guideline. Mark the dart notches and the dart apex.

Then shift the pattern in another $1/16"$ maintaining shoulder level, so that the pattern is now a total of $3/16"$ from the length guideline. Trace the remainder of the shoulder to and around the armhole intersection. Blend the shoulder seam as shown on page 19.

5 ARMHOLE GRADE

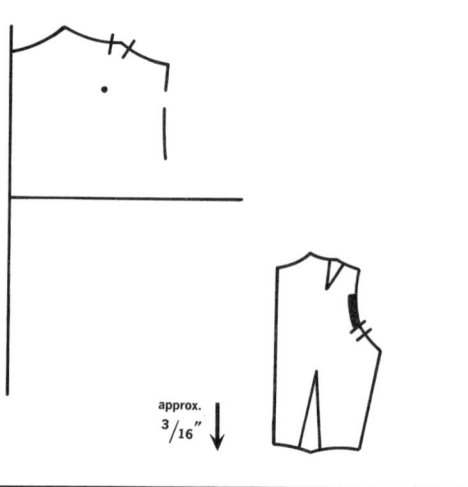

7 SIDE SEAM GRADE

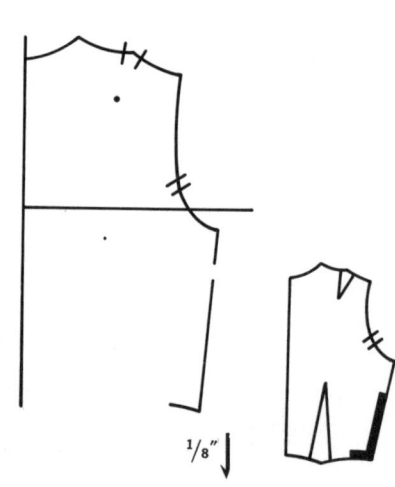

Shift the pattern down $1/8"$ from its previous position. The cross control line of the pattern should now be a total of $1/4"$ below the cross guideline on the paper. Trace the side seam to and around the waistline intersection.

8 APEX WIDTH GRADE

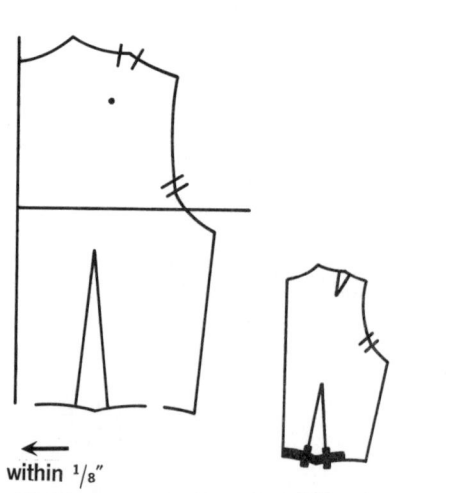

1. In all steps except 4* and 6*, the measurements used are standard from one size to the next no matter if it be a 1", 1½", or 2" grade. Refer to the bodice grading chart for the measurements for steps 4 and 6 *when doing other than the 1½" grade shown below*, and follow directions given for the front bodice.

2. The smaller diagram in the lower corner of each box depicts the pattern being graded, with the section to be traced in that step marked in heavy lines.

Shift the pattern up on the length guideline until the cross control line of the pattern is ⅛" above the cross guideline on the paper. Trace the middle of the shoulder in order to establish the shoulder level.

3 NECK GRADE

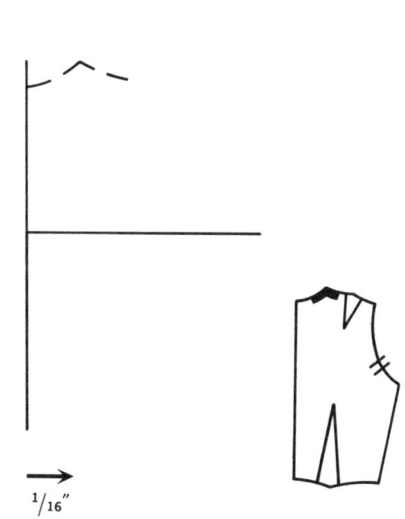

Shift the pattern in 1/16" from the length guideline. Before tracing, adjust the pattern so that the shoulder seam is once again directly on the shoulder level traced in step 2. This adjustment is known as "maintaining shoulder level". Now trace the shoulder-neckline intersection. Blend the neckline as shown on page 20.

Shift the pattern down so that the cross control line of the pattern is ⅛" below the cross guideline on the paper. The total downward shift of the pattern from its previous position will be approximately 3/16". Trace the middle of the armhole.

6 CROSS BUST GRADE*

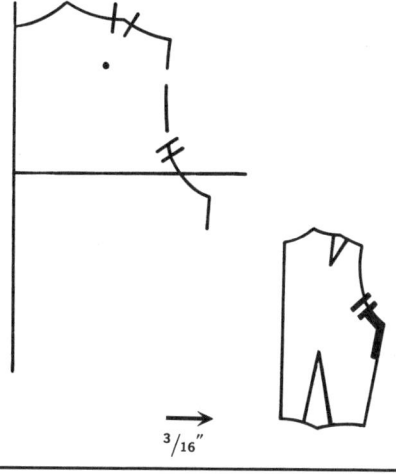

Shift the pattern in 3/16" from its previous position. The pattern should now be a total of ⅜" from the length guideline. Trace the lower armhole and side seam intersection. Blend the armhole as shown on page 21. Mark the armhole notches.

Shift the pattern out until it is within ⅛" of the length guideline. Trace the waistline and mark the dart notches. Blend the waistline across to the length guideline, thus establishing the new waistline-Center Back intersection. (See page 19).

9 DART LENGTH GRADE

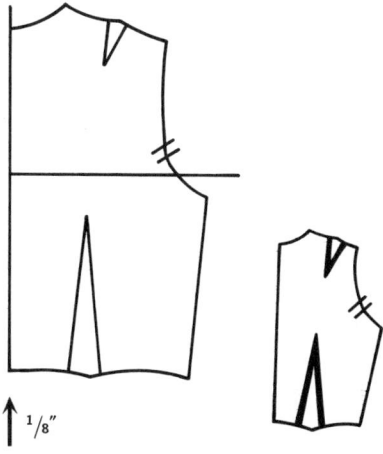

With the pattern still within ⅛" of the length guideline, shift the pattern up ⅛" from its previous position and mark the dart apex. Lift the pattern off the paper and draw in the dartlines. The back bodice 1½" grade is now complete.

HOW THE BODY GROWS

BACK BODICE WITH NECK DART

1½" Grade

1. The back bodice with the neck dart grades the same as the front bodice in its areas of body growth.

2. The back neck dart and back shoulder dart do not grade in length. They shape a general body area rather than a specific point, and they would become unnecessarily long and narrow if graded in length.

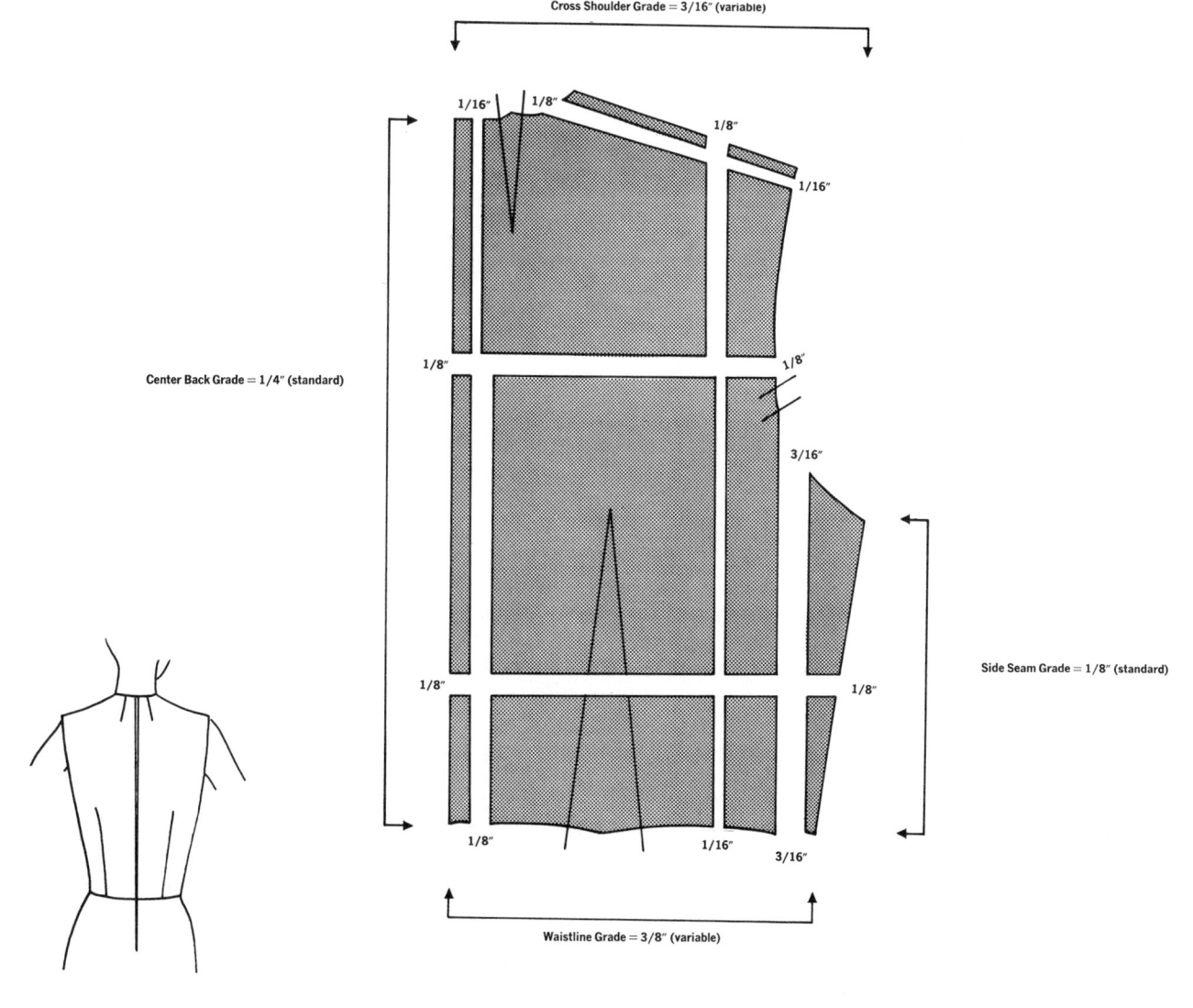

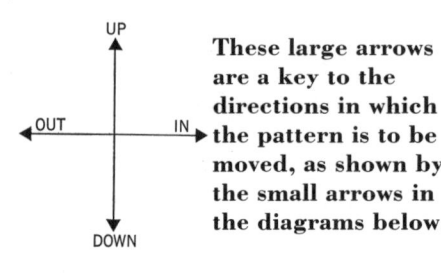

These large arrows are a key to the directions in which the pattern is to be moved, as shown by the small arrows in the diagrams below.

DIRECTIONS FOR A BACK BODICE 1½ INCH GRADE
(NECK DART)

1 CENTER BACK

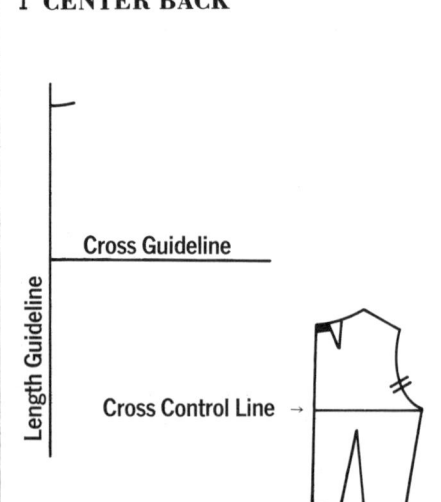

Draw a cross control line on the pattern by squaring a line from Center Back to the underarm-side seam intersection. On paper, draw a length guideline, and square a cross guideline from it. Place Center Back of the pattern on the length guideline, with the cross control line of the pattern on the cross guideline. Trace the neckline-Center Back intersection. (The length guideline will be the Center Back of the graded pattern).

2 SHOULDER LEVEL GRADE

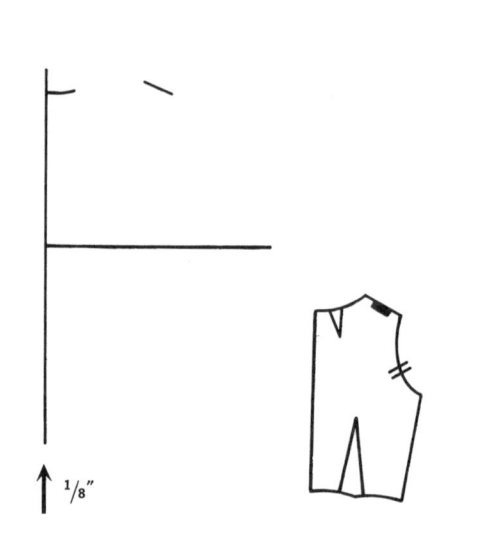

4 CROSS SHOULDER GRADE*

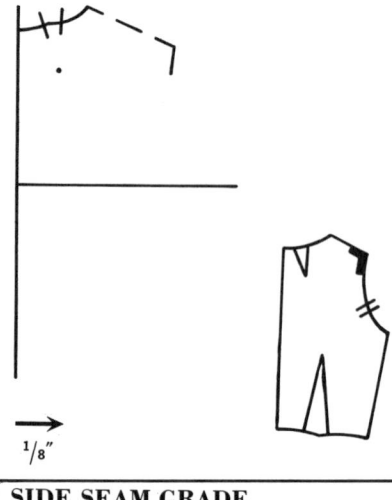

Shift the pattern in ⅛″ from its previous position, maintaining shoulder level. The pattern should now be a total of ³⁄₁₆″ from the length guideline. Trace the remainder of the shoulder, and the armhole intersection. Blend the shoulder seam as shown on page 19.

5 ARMHOLE GRADE

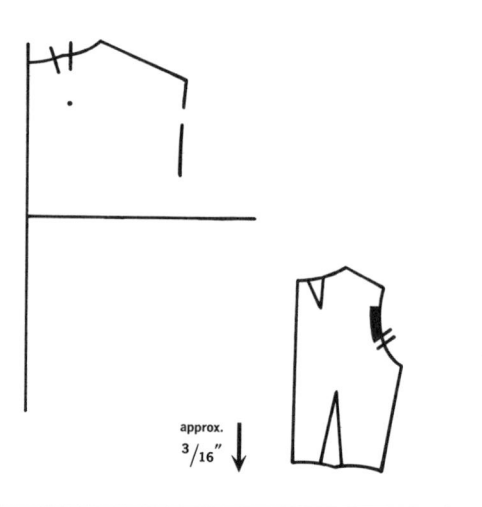

7 SIDE SEAM GRADE

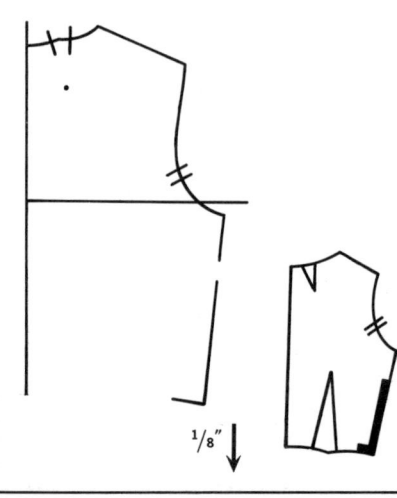

Shift the pattern down ⅛″ from its previous position. The cross control line of the pattern should now be a total of ¼″ below the cross guideline on the paper. Trace the side seam to and around the waistline intersection.

8 APEX WIDTH GRADE

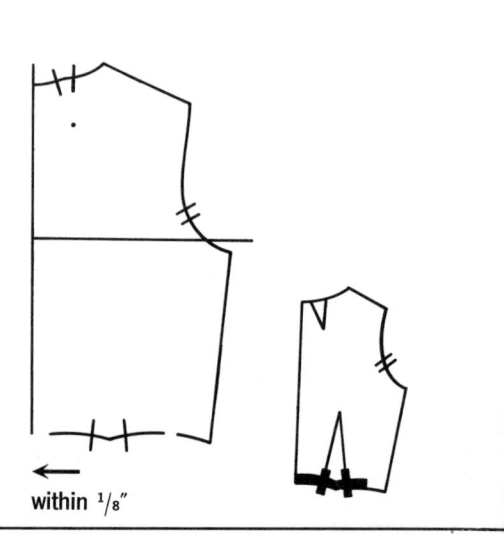

34

1. In all steps except 4* and 6*, the measurements used are standard from one size to the next no matter if it be a 1″, 1½″, or 2″ grade. Refer to the bodice grading chart for the measurements for steps 4 and 6 *when doing other* than the 1½″ grade shown below, and follow directions given for the front bodice.

2. The smaller diagram in the lower corner of each box depicts the pattern being graded, with the section to be traced in that step marked in heavy lines.

Shift the pattern up on the length guideline until the cross control line of the pattern is ⅛″ above the cross guideline on the paper. Trace the middle of the shoulder in order to establish the shoulder level.

3 NECK GRADE

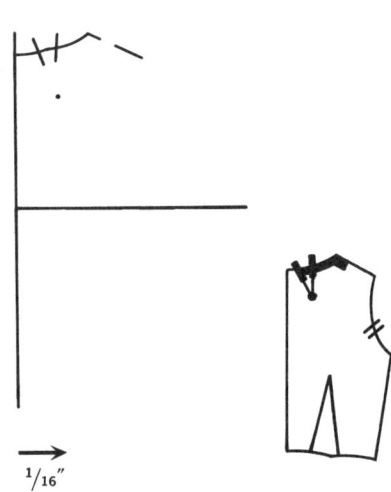

Shift the pattern in 1/16″ from the length guideline. Before tracing, adjust the pattern so that the shoulder seam is once again directly on the shoulder level traced in step 2. This adjustment is known as "maintaining shoulder level". Now trace the shoulder-neckline intersection and mark the neck dart notches and the dart apex. Blend the neckline as shown on page 20.

Shift the pattern down so that the cross control line of the pattern is ⅛″ below the cross guideline on the paper. The total downward shift of the pattern from its previous position will be approximately 3/16″. Trace the middle of the armhole.

6 CROSS BUST GRADE*

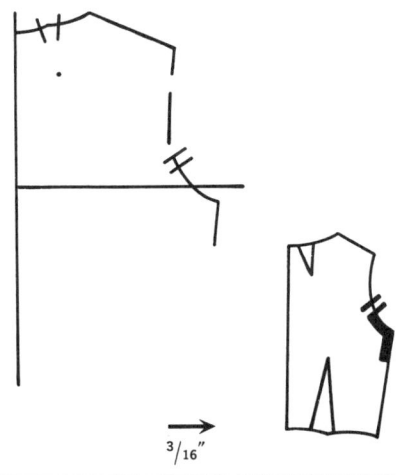

Shift the pattern in 3/16″ from its previous position. The pattern should now be a total of ⅜″ from the length guideline. Trace the lower armhole and the side seam intersection. Blend the armhole as shown on page 21. Mark the armhole notches.

Shift the pattern out until it is within ⅛″ of the length guideline. Trace the waistline and mark the dart notches. Blend the waistline across to the length guideline, thus establishing the new waistline-Center Back intersection. (See page 19).

9 DART LENGTH GRADE

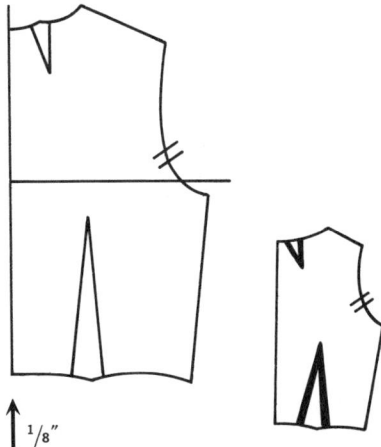

With the pattern still within ⅛″ of the length guideline, shift the pattern up ⅛″ from its previous position and mark the dart apex. Lift the pattern off the paper and draw in the dartlines. The back bodice 1½″ grade is now complete.

HOW THE BODY GROWS

SKIRT
1½" Grade
1. When grading separate skirts, the waistline increase is divided equally between the darts.
2. When the skirt is part of an entire garment, the waistline increase is always the same as that of the bodice waistline, and the distribution of the skirt darts should match the bodice dart lines.
3. The total of the width grades widen the waistline, hipline, and sweep of the skirt.
4. The lengthwise increases of ¼" per size may be held after Misses size 14, and Junior size 13.
5. A skirt waistband is increased the same amount as the waistline of the skirt. The width of the waistband remains the same from size to size.

COLLAR
1½" Grade
1. The neckline of the collar must always be graded in front and back the same amount as the neckline of the bodice to which it will be attached.
2. The collar usually is not graded in width in order to retain its desired proportions.

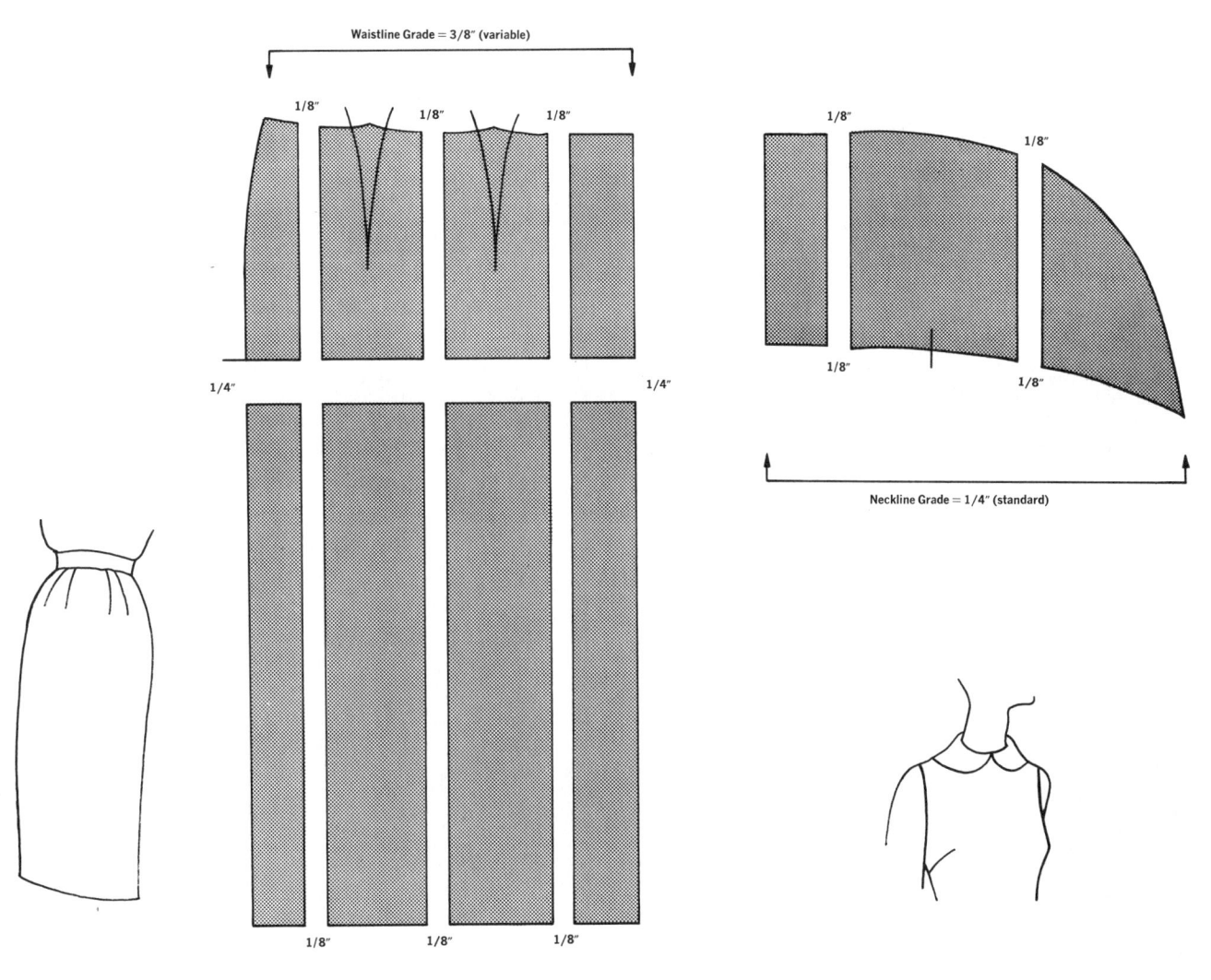

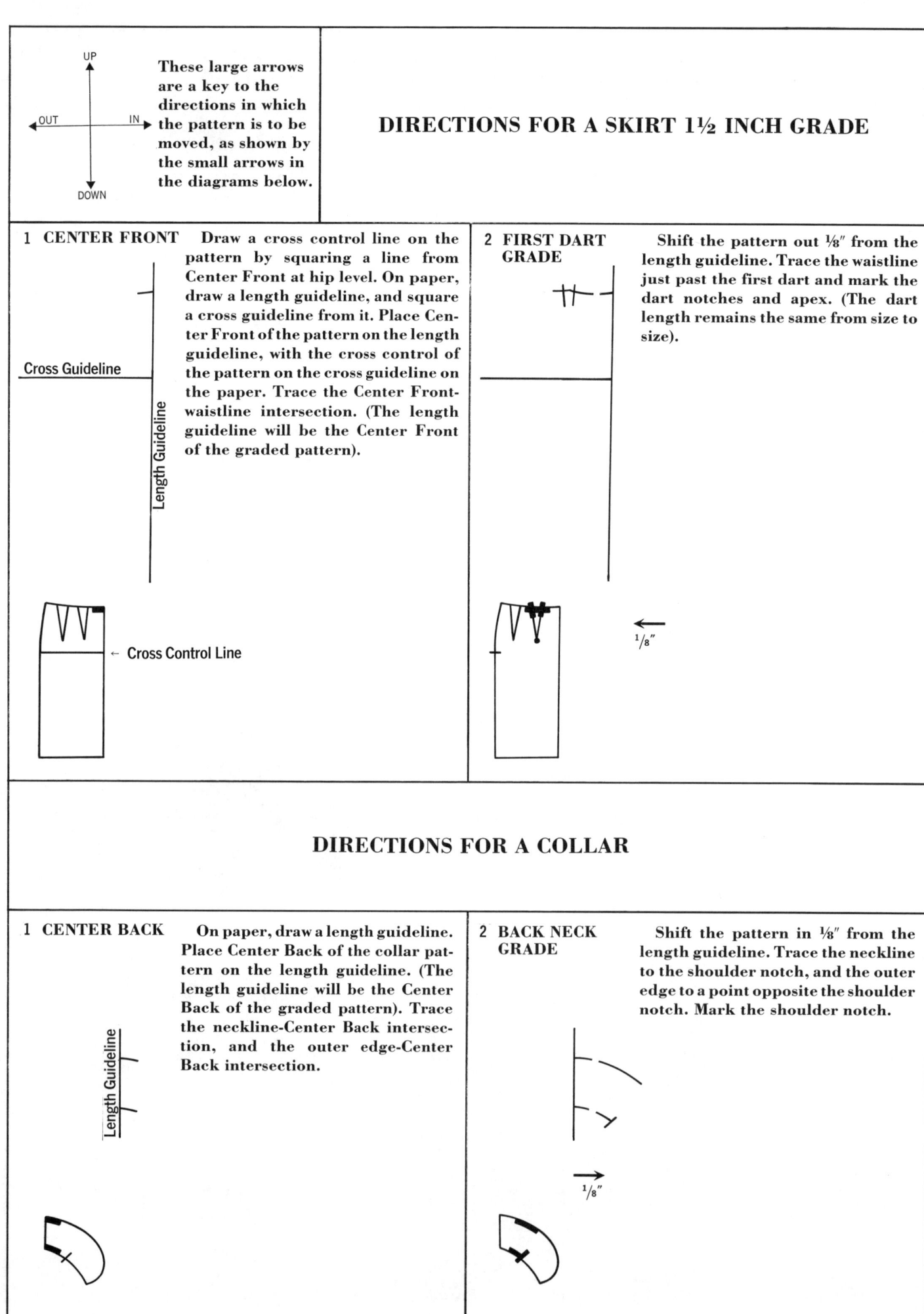

These large arrows are a key to the directions in which the pattern is to be moved, as shown by the small arrows in the diagrams below.

DIRECTIONS FOR A SKIRT 1½ INCH GRADE

1 CENTER FRONT Draw a cross control line on the pattern by squaring a line from Center Front at hip level. On paper, draw a length guideline, and square a cross guideline from it. Place Center Front of the pattern on the length guideline, with the cross control of the pattern on the cross guideline on the paper. Trace the Center Front-waistline intersection. (The length guideline will be the Center Front of the graded pattern).

2 FIRST DART GRADE Shift the pattern out ⅛" from the length guideline. Trace the waistline just past the first dart and mark the dart notches and apex. (The dart length remains the same from size to size).

DIRECTIONS FOR A COLLAR

1 CENTER BACK On paper, draw a length guideline. Place Center Back of the collar pattern on the length guideline. (The length guideline will be the Center Back of the graded pattern). Trace the neckline-Center Back intersection, and the outer edge-Center Back intersection.

2 BACK NECK GRADE Shift the pattern in ⅛" from the length guideline. Trace the neckline to the shoulder notch, and the outer edge to a point opposite the shoulder notch. Mark the shoulder notch.

1. In all steps except 3* for the skirt grade, the measurements used are standard from one size to the next no matter if it be a 1″, 1½″, or 2″ grade. Refer to the skirt grading chart for the measurements for step 3 *when doing other than* the 1½″ grade shown below. Subtract ⅛″ from column 7, divide the remainder in half, and use that half in each shift of step 3.

2. The smaller diagram in the lower corner of each box depicts the pattern being graded, with the section to be traced in that step marked in heavy lines.

3 SECOND DART GRADE*

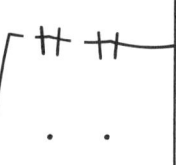

Shift the pattern out ⅛″ from its previous position. The pattern should now be a total of ¼″ from the length guideline. Trace the waistline just past the second dart and mark the dart notches and apex.

Then shift the pattern out another ⅛″. The pattern should now be a total of ⅜″ from the length guideline. Trace the waistline to and around the side seam intersection, and down the side seam to the hip level. Mark the hip notch. (For a skirt with one dart, the ¼″ waistline grade, done in two shifts in this step, is done in one ¼″ shift in this step).

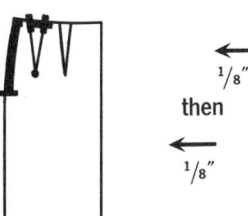

← ⅛″
then
← ⅛″

4 SIDE SEAM GRADE

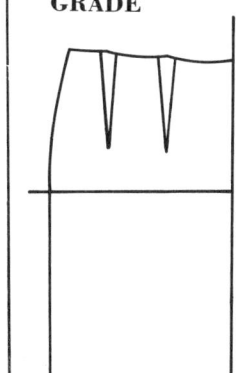

Shift the pattern down so that the cross control line of the pattern is ¼″ below the cross guideline on the paper. Trace the side seam to and around the hemline intersection, and blend across to the length guideline. Draw in the dartlines.

The back skirt is graded in a similar manner.

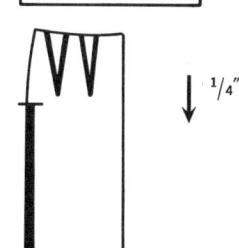

↓ ¼″

1½ INCH GRADE

3 FRONT NECK GRADE

Shift the pattern in ⅛″ from its previous position. The pattern should now be a total of ¼″ from the length guideline. Trace the remainder of the collar neckline and the outer edge.

→ ⅛″

4 OTHER TYPES OF COLLARS

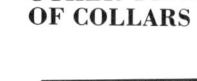
Convertible

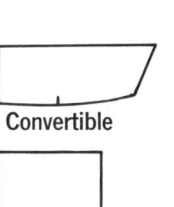
Sailor

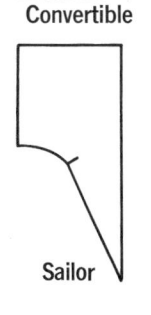
Mandarin

Follow the same procedures for all separate collars, such as the buster brown and eton. The width of such collars remains the same from size to size, as a general rule.

HOW THE BODY GROWS

TORSO | **1½″ Grade**

1. The side bust dart is graded ⅛″ per size, even though that area of the body on a 1½″ grade grows larger than that per size. This in effect foreshortens the dart, allowing it to cup more of the fuller bust.

2. When the torso sloper is used as the basis for overblouse or tuck-in blouse designs, no length grade need be given in the hip section. When the torso sloper is used as the basis for a one-piece dress, the length grade may be given in the hip section as it is for skirts.

3. Dart pick-up remains the same from size to size, since the areas that shape the dart at its widest and narrowest ends grow in proportion to each other, thus allowing the dart pick-up to remain the same.

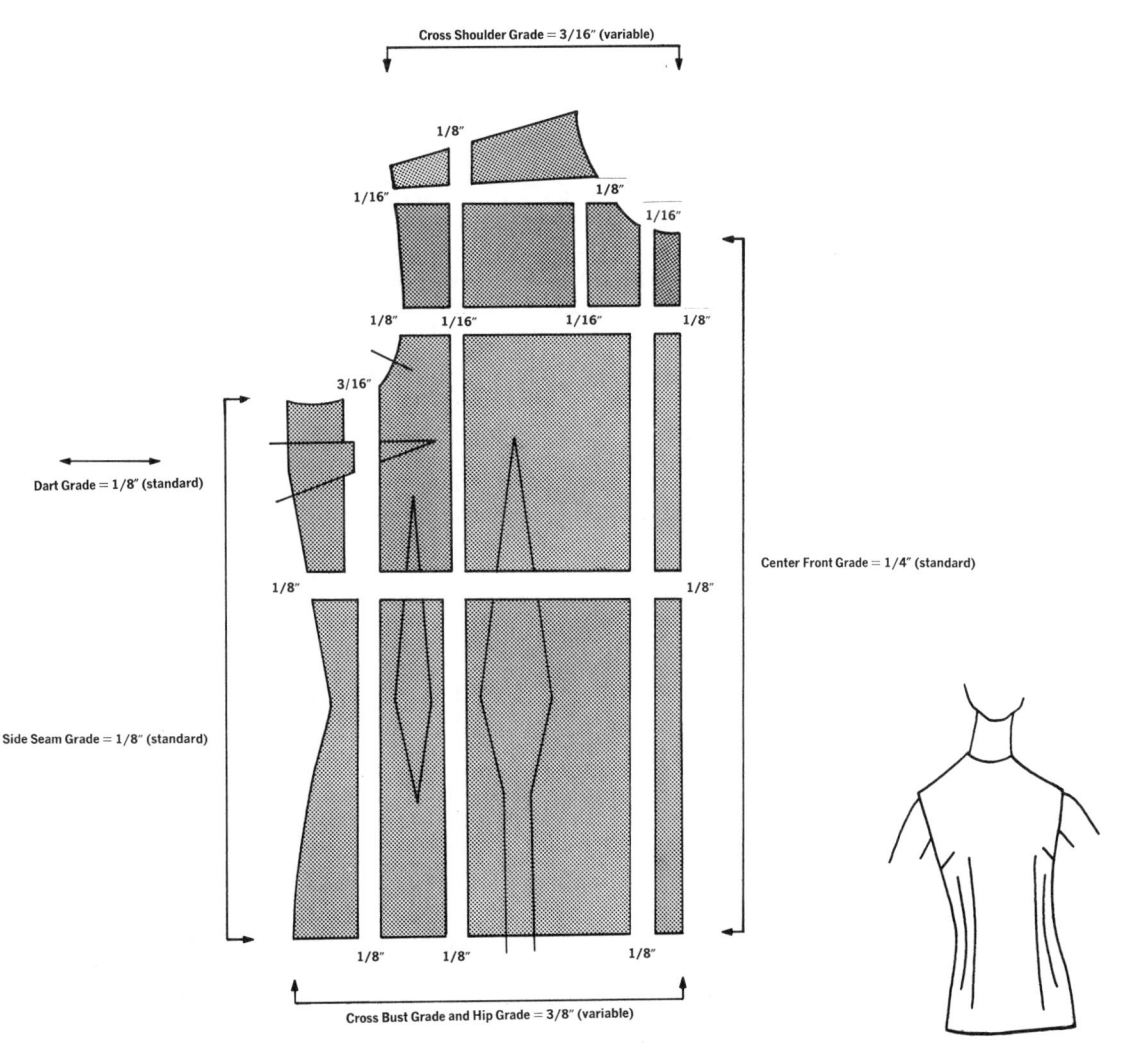

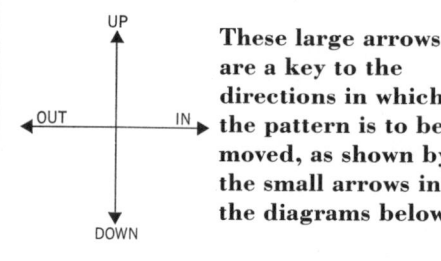

These large arrows are a key to the directions in which the pattern is to be moved, as shown by the small arrows in the diagrams below.

DIRECTIONS FOR A FRONT TORSO 1½″ GRADE

1 to 6 UPPER TORSO GRADE*

Follow steps 1 through 6 of the Front Bodice on pages 26 and 27. In step 6, trace the side seam to just past the side bust dart notches, and mark the dart notches.

7 SIDE SEAM GRADE

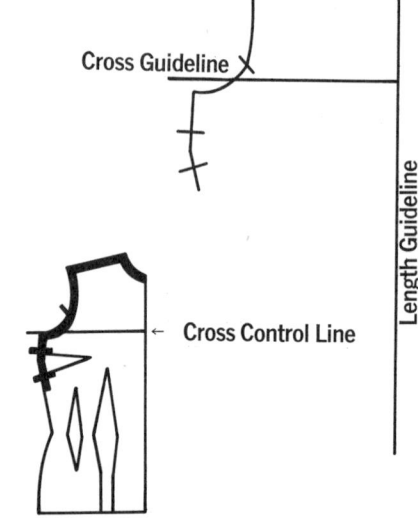

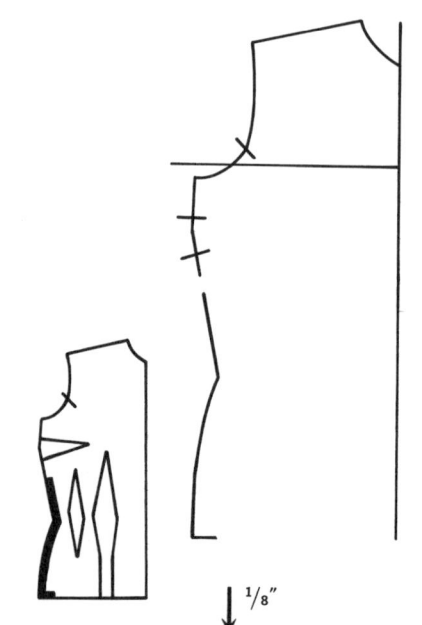

9 WAISTLINE DARTS—II

Shift the pattern up ⅛″ from its previous position. The control line of the pattern should now be a total of ⅛″ below the cross guideline on the paper. Mark the apex of the side waistline dart, and the apex of the side bust dart.

Then shift the pattern in until it is within ⅛″ of the length guideline. Mark the apex of the bust waistline dart.

10 WAISTLINE DARTS—III

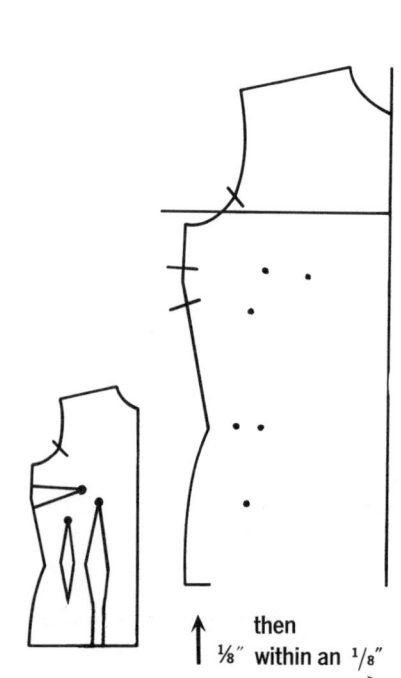

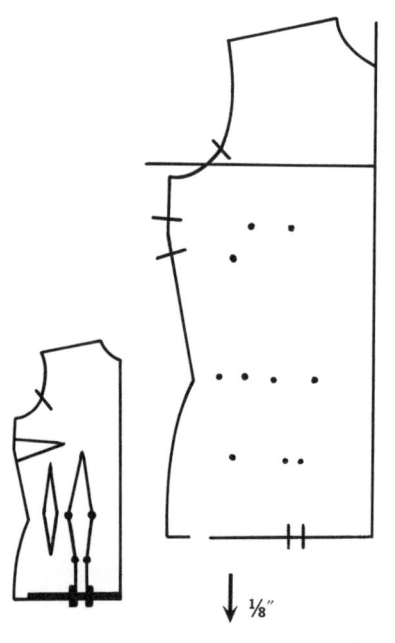

1. In all steps except 4* and 6*, the measurements used are standard from one size to the next, no matter if it be a 1", 1½", or 2" grade. Refer to the bodice grading chart for the measurements for steps 4 and 6 *when doing other than the 1½" grade shown below*, and follow the directions given for the Front Bodice. In step 8* shift in an amount that will distribute the waistline increase equally between the waistline darts and the side seam.

2. The smaller diagram in the lower corner of each box depicts the pattern being graded, with the section to be traced in that step marked in heavy lines.

Shift the pattern down ⅛" from its previous position. The control line of the pattern should now be a total of ¼" below the cross guideline on the paper. Trace the side seam past the waistline, to and around the side seam—hipline intersection.

8 WAISTLINE DARTS—I*

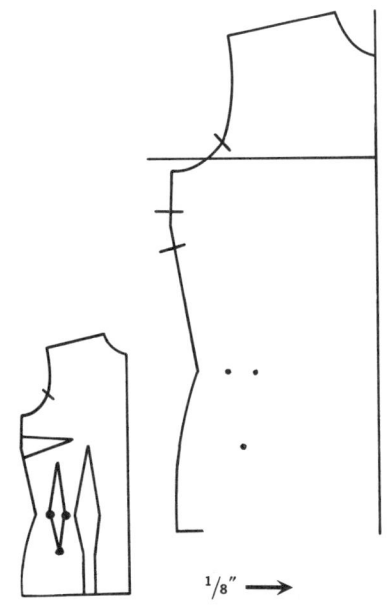

⅛" →

Shift the pattern in ⅛" from its previous position. On a 1½" grade, the pattern should now be a total of ¼" from the length guideline. Mark the hip level apex and waistline pick-up of the side waistline dart.

For a 1" grade, shift the pattern in 1/16".

For a 2" grade, shift the pattern in 3/16".

Shift the pattern down ⅛" from its previous position. The control line of the pattern should now be a total of ¼" below the cross guideline on the paper. Mark the waistline pick-up of the bust waistline dart, and mark the hip level apexes and crossmarks of the bust waistline dart. Trace the hipline from the side seam intersection to Center Front, and blend across to the length guideline, thus establishing the new hipline—Center Front intersection.

11 DART COMPLETION

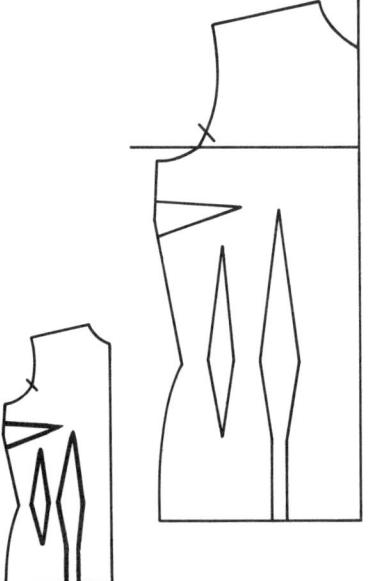

Lift the pattern off the paper and draw in all the dartlines. The front torso 1½" grade is now complete. For the back torso grade, follow the same procedure, using steps 1 through 6 of the back bodice lesson and continuing with steps 7 through 10 of the front torso lesson.

HOW THE BODY GROWS

SLACKS

1½″ Grade

1. When grading slacks, the waistline increase is divided equally between the darts, in a similar manner to the basic skirt method.

2. The crotch is the most important area to consider in order to have proper fit. The crotch length increase of ¼″ per size is never held, as it may be in skirts. The crotch width increase is ⅛″ per size.

3. The leg is graded ⅛″ per size for added length. In grading shorts, that ⅛″ grade is not given.

4. On tapered slacks, the ankle is always graded one-half of the waistline grade. For example, the split diagram shows the 1½″ grade with the waistline increasing ⅜″; thus the ankle increases 3/16″. On straight slacks, and flared slacks such as bell bottoms, the ankle is graded the same amount as the waist.

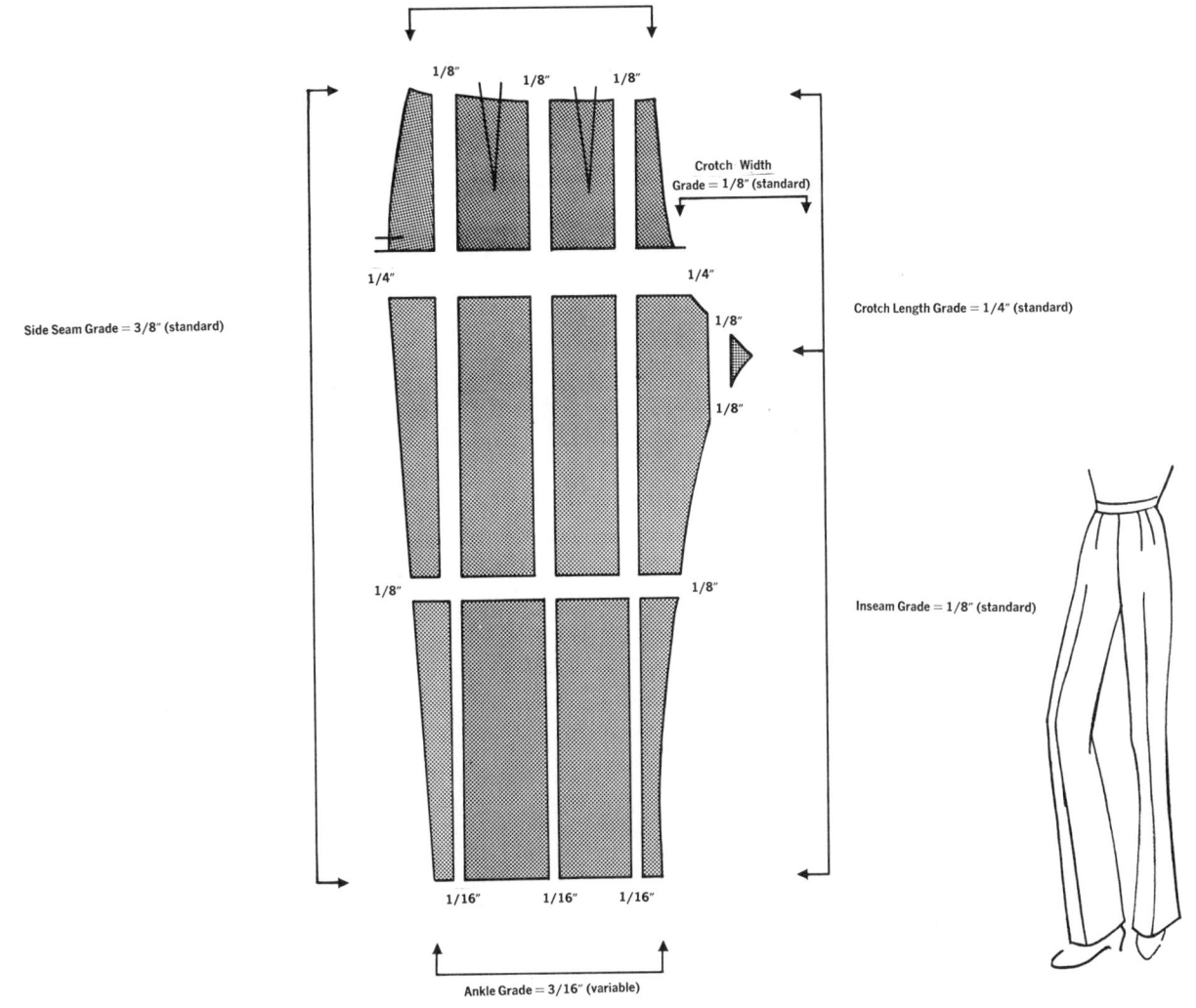

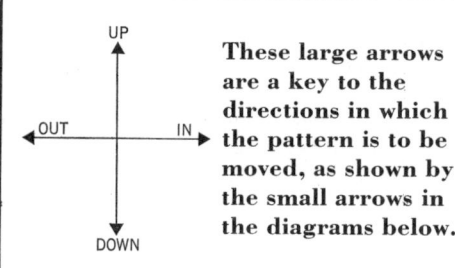

These large arrows are a key to the directions in which the pattern is to be moved, as shown by the small arrows in the diagrams below.

DIRECTIONS FOR A SLACK 1½ INCH GRADE

1 CENTER FRONT

Draw a cross control line on the pattern at crotch level at a right angle to the grainline. On paper, draw a length guideline, and square a cross guideline from it. Place the pattern on the paper so that the cross control line lies on the cross guideline, and the crotch point is touching the length guideline on the paper. Trace the crotch curve halfway up to the hip level.

2 CROTCH GRADE

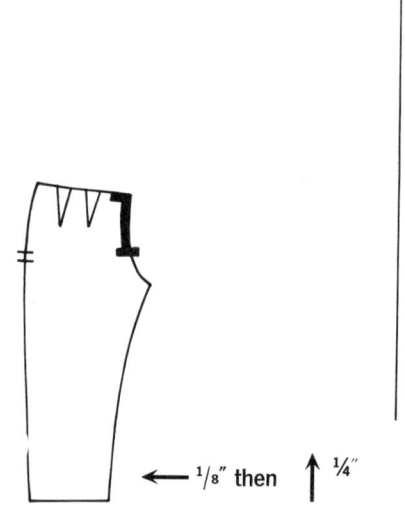

←— ⅛″ then ↑ ¼″

4 SIDE SEAM*

Shift the pattern out ⅛″ from its previous position. The crotch point of the pattern should now be a total of ½″ from the length guideline on the paper. Trace the waistline to and around the side seam intersection, and down the side seam to the hip level. Mark the hip notches.

Then shift the pattern down ¼″ from its previous position so that the control line of the pattern is once again on the cross guideline. Trace the side seam to the crotch level.

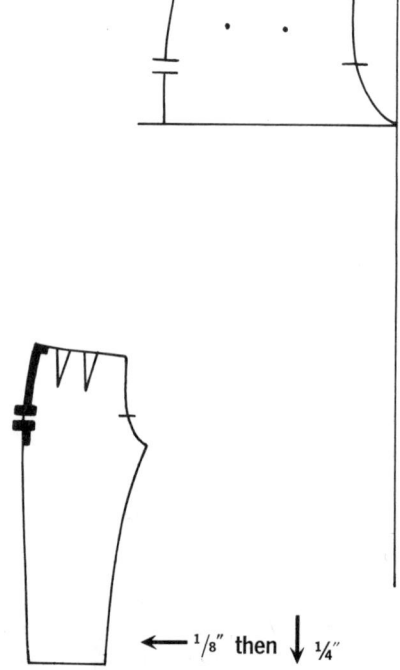

←— ⅛″ then ↓ ¼″

5 LENGTH GRADE

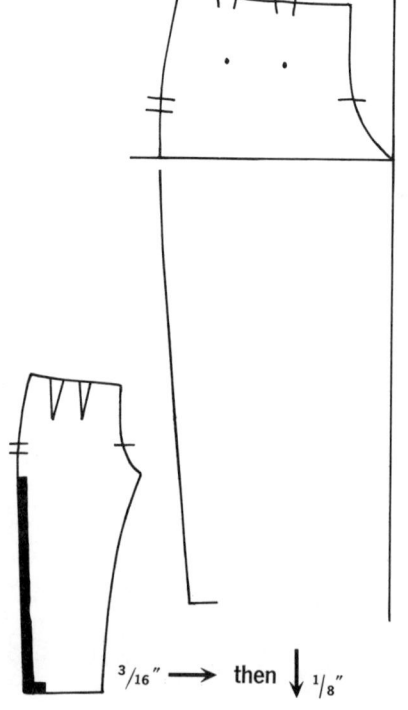

³⁄₁₆″ —→ then ↓ ⅛″

46

1. In all steps except the second shift in step 3* and the first shift in step 4*, the measurements used are standard from one size to the next no matter if it be a 1", 1½", or 2" grade. Refer to the skirt grading chart for the measurements of the second shift in step 3 and the first shift in step 4 *when doing other* than the 1½" grade shown below. Subtract ⅛" from column 7, divide the remainder in half, and use that half in the second shift of step 3 and the first shift of step 4.

2. The smaller diagram in the lower corner of each box depicts the pattern being graded, with the section to be traced in that step marked in heavy lines.

Shift the pattern out so that the crotch point of the pattern is ⅛" from the length guideline on the paper. Trace nothing.

Then shift the pattern up so that the cross control line is ¼" above the cross guideline. Trace Center Front and the waistline intersection. Blend the crotch seam and mark the crotch crossmark.

3 WAISTLINE GRADE*

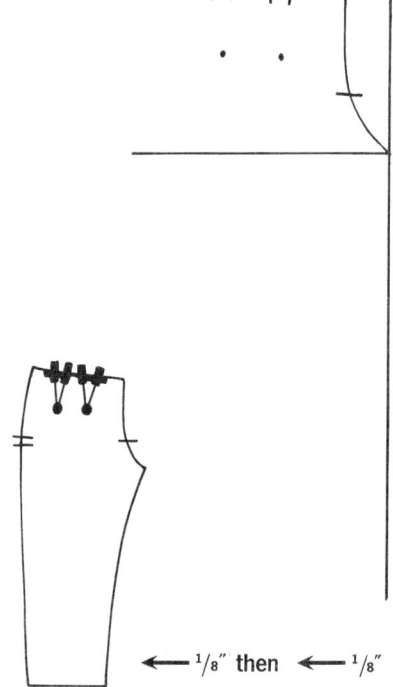

← ⅛" then ← ⅛"

Shift the pattern out ⅛" from its previous position. The crotch point of the pattern should now be a total of ¼" from the length guideline on the paper. Trace the first dart notches and mark the dart apex.

Then shift the pattern out another ⅛" from its previous position. The crotch point of the pattern should now be a total of ⅜" from the length guideline on the paper. Trace the second dart notches and mark the dart apex.

Shift the pattern in ³⁄₁₆" from its previous position. The crotch point of the pattern should now be ⁵⁄₁₆" from the length guideline. Trace nothing. For a 1" grade, shift in ⅛". For a 2" grade, shift in ¼".

Then shift the pattern down so that the cross control line of the pattern is ⅛" below the cross guideline on the paper. Trace the side seam ankle intersection. Blend the side seam from the crotch level to the ankle. Note: On a straight slack, omit the first shift above.

6 ANKLE GRADE

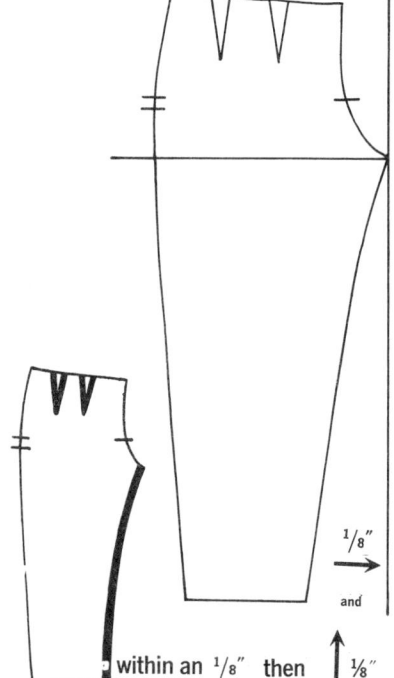

within an ⅛" then ↑ ⅛" ⅛" → and

Shift the pattern in from its previous position so that the crotch point of the pattern is within ⅛" of the length guideline. Trace the ankle line to and around the inseam intersection.

Then shift the pattern up ⅛", and in ⅛", so that the crotch point is back to its original starting position, and blend the inseam from the ankle intersection to the crotch. Lift the pattern off the paper and draw in the dart lines.

The back slack pattern is graded in a similar manner.

HOW THE BODY GROWS

SET-IN SLEEVE

1½" Grade

1. The cap of the sleeve is always graded in relation to the corresponding bodice armhole.

2. The elbow width is always graded the same amount as the bicep width.

3. The elbow dart is not graded in width, and is increased in length an amount equal to one-fourth of the entire elbow grade, as shown in column 4 of the sleeve grading chart.

4. A fitted sleeve wrist is graded one-half of the bicep grade, while on a straight sleeve the wrist is graded the same as the bicep.

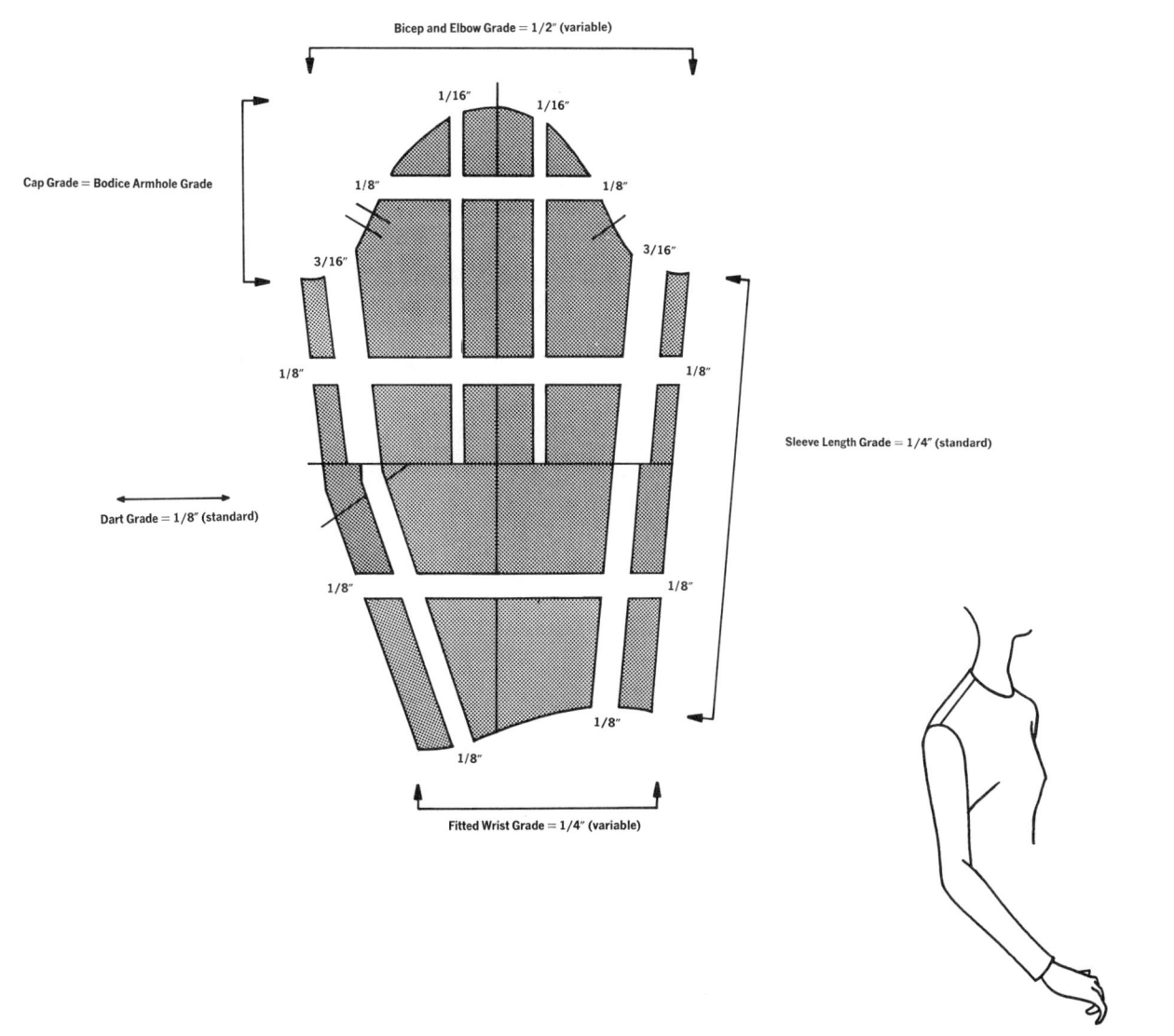

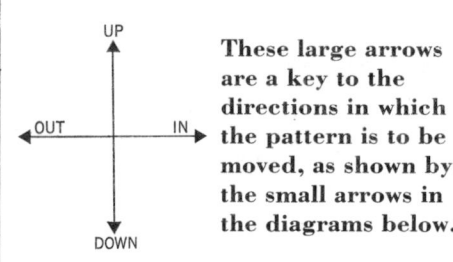

These large arrows are a key to the directions in which the pattern is to be moved, as shown by the small arrows in the diagrams below.

DIRECTIONS FOR A SET-IN SLEEVE 1½" GRADE

1 BICEP GRADE*

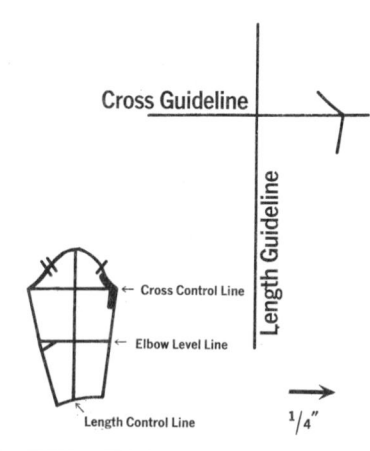

Draw a cross control line on the pattern by ruling a line across the bicep. Draw an elbow level line on the pattern. Draw a length control line on the pattern by ruling a line through the center of the sleeve. On paper, draw a length guideline and square a cross guideline from it. Place the pattern with its control lines on top of the corresponding guidelines on the paper. Shift the pattern in ¼" from the length guideline. Trace the front bicep corner.

2 FRONT CAP GRADE

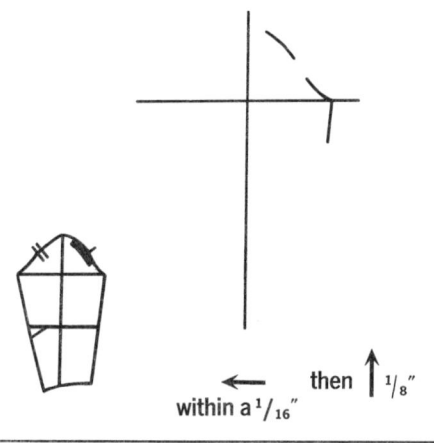

4 CAP WIDTH GRADE—back

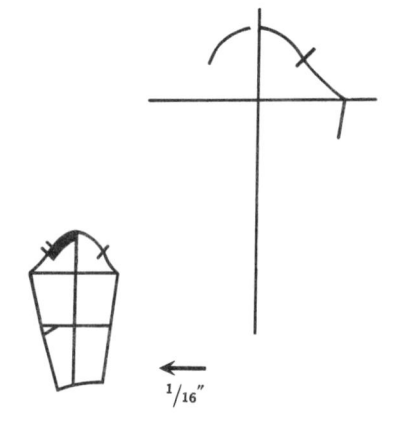

Shift the pattern out 1/16" from the length guideline. Trace the top and middle of the cap.

5 BACK CAP GRADE*

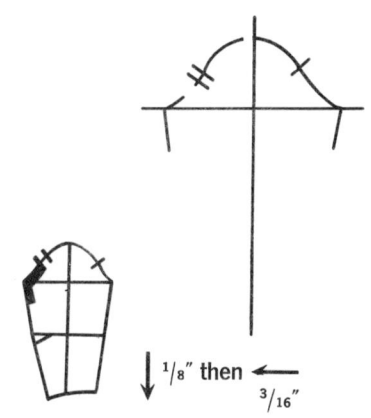

7 BACK WRIST GRADE

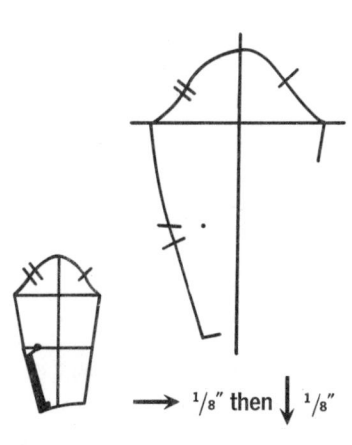

Shift the pattern in ⅛". On a 1½" grade, the length control line of the pattern should now be ⅛" from the length guideline on the paper. Mark the dart apex. (On all grades, the dart should grow in length an amount equal to one-fourth of the total elbow grade. See column 4 on the sleeve grading chart).

Then shift the pattern down ⅛" from its previous position. The cross control line of the pattern should now be a total of ¼" below the cross guideline on the paper. Trace the back wrist intersection, and blend the back underarm seam from the elbow dart to the wrist intersection.

8 FRONT WRIST GRADE

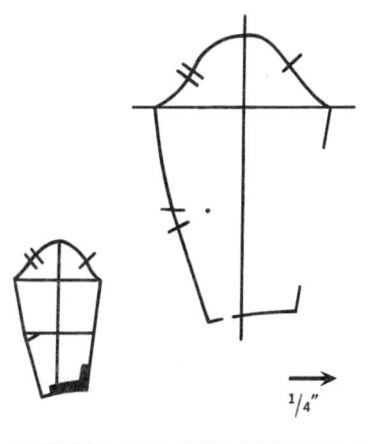

1. In all steps except 1* and the second shift of step 5*, the measurements used are standard from one size to the next, no matter if it be a 1″, 1½″, or 2″ grade. Refer to column 1 of the sleeve grading chart for the measurements for steps 1 and the second shift of step 5 *when doing* other than the 1½″ grade shown below. For step 1, divide the measurement in half before using. For the second shift of step 5 divide the measurement in half and subtract 1/16″ from that measurement before using.

2. The smaller diagram in the lower corner of each box depicts the pattern being graded, with the section to be traced in that step marked in heavy lines.

Shift the pattern out until the length control line of the pattern is within 1/16″ of the length guideline on the paper. Trace nothing.

Then shift the pattern up until the cross control line of the pattern is 1/8″ above the cross guideline on the paper. Trace the middle of the cap.

3 CAP WIDTH GRADE—front

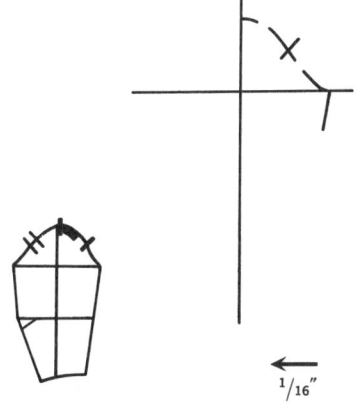

← 1/16″

Shift the pattern out 1/16″ until the length control line of the pattern lies on top of the length guideline on the paper. Trace the top of the cap, and mark the center cap notch. Blend the sleeve cap and mark the front cap notch.

Shift the pattern down 1/8″ until the cross control line of the pattern lies on top of the cross guideline on the paper. Trace nothing.

Then shift the pattern out 3/16″ from its previous position. On a 1½″ grade, the length control line of the pattern should now be a total of 1/4″ out from the length guideline on the paper. Trace the back bicep corner. Blend the sleeve cap and mark the back cap notches.

6 UNDERARM GRADE—Back

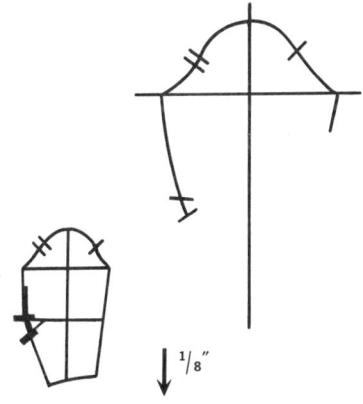

↓ 1/8″

Shift the pattern down until the cross control line of the pattern is 1/8″ below the cross guideline on the paper. Trace the back underarm seam through the elbow dart. Mark the elbow dart notches.

Shift the pattern in 1/4″ from its previous position. The length control line of the pattern should now be 1/8″ in from the length guideline on the paper. Trace the wrist to and around the front underarm seam intersection. Blend the wrist line.

9 UNDERARM GRADE—Front

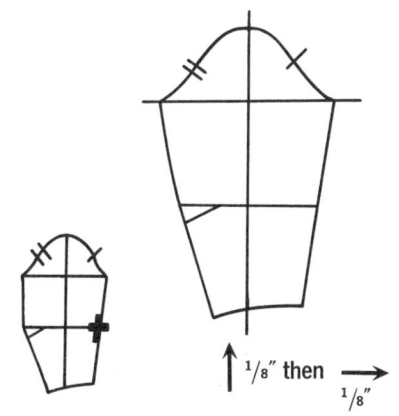

↑ 1/8″ then → 1/8″

Shift the pattern up 1/8″ until the cross control line of the pattern is 1/8″ below the cross guideline on the paper. Trace nothing.

Then shift the pattern in 1/8″ from its previous position. The length control line of the pattern should now be a total of 1/4″ in from the length guideline on the paper. Mark the elbow level crossmark. Blend a line from the bicep corner to the wrist, using the elbow crossmark as a guide. Draw in the dart lines. The set-in sleeve 1½″ grade is now complete.

PART 3

Grading Intermediate Designs

BODICE & YOKE

BODICE & MIDRIFF

PRINCESS BODICE

SIX-GORE SKIRT

CIRCLE SKIRT

SHAWL COLLAR

HOW THE BODY GROWS

BODICE AND YOKE

1½" Grade

1. The yoke section is not graded in length; the bodice section receives the entire length grade. This practice of holding the yoke in length maintains the neckline-to-yokeseam proportion as created by the designer.

2. The shirring distance between the notches is not graded. This too maintains the original shirring proportions.

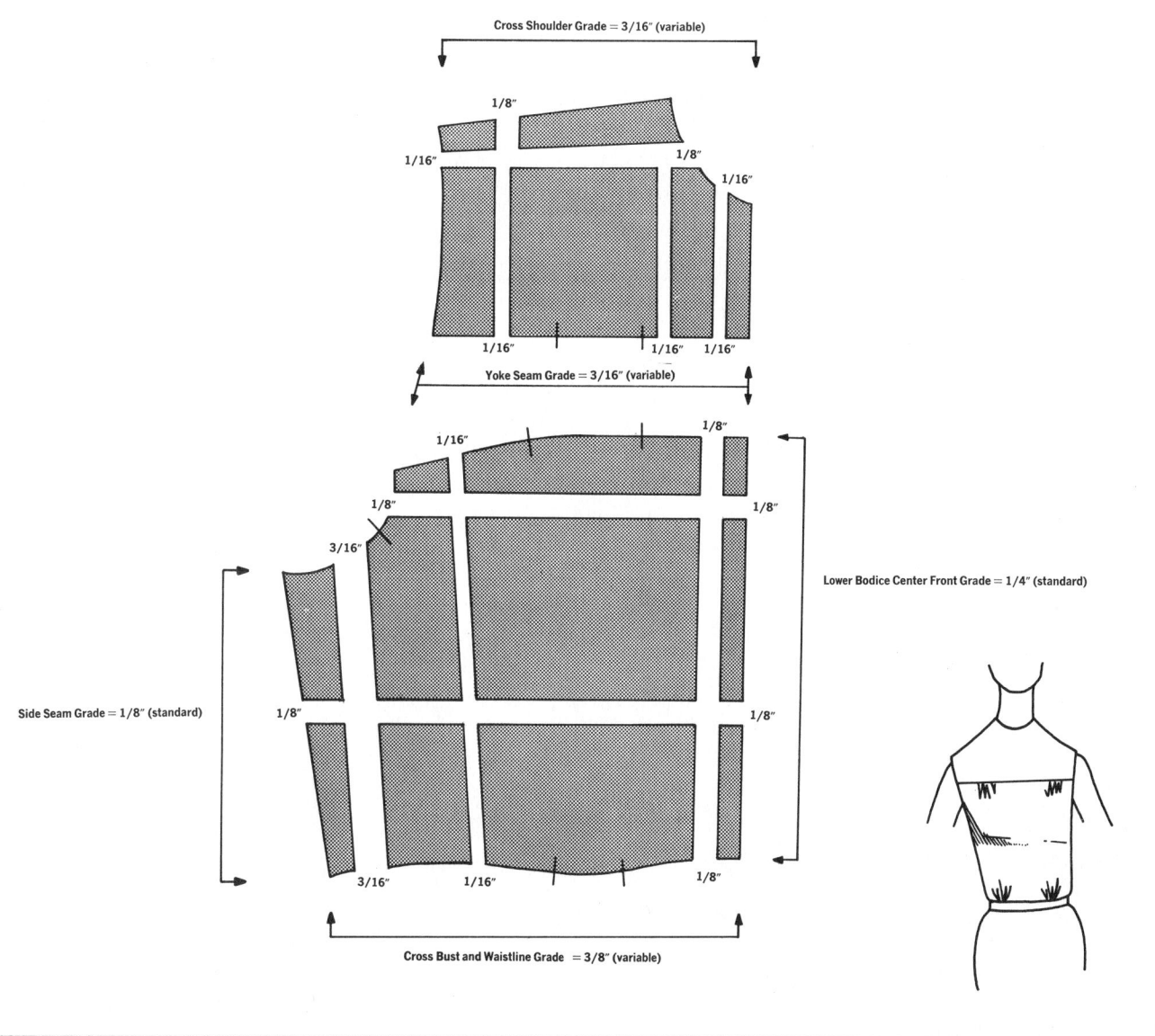

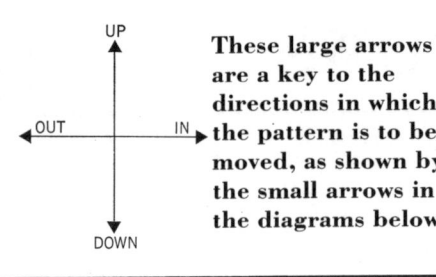

These large arrows are a key to the directions in which the pattern is to be moved, as shown by the small arrows in the diagrams below.

DIRECTIONS FOR A FRONT BODICE AND YOKE 1½" GRADE

1 to 4   YOKE GRADE*

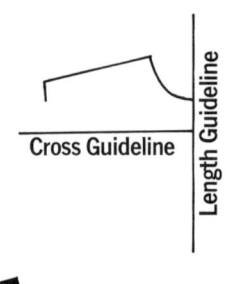

Draw a cross control line on the yoke pattern by squaring a line halfway down Center Front. On paper, draw a length guideline and square a cross guideline from it. Place Center Front of the pattern on the length guideline, with the cross control line of the pattern on the cross guideline. Then follow steps 1 through 4 of the front bodice grade, pages 26 and 27.

7   BODICE GRADE

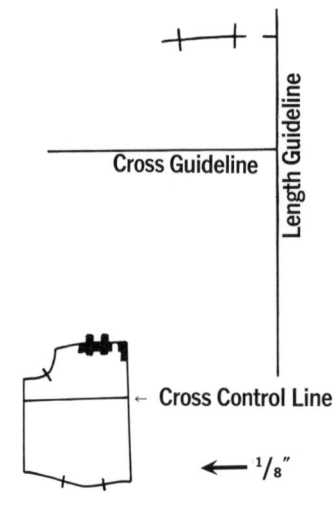

Draw a cross control line on the pattern by squaring a line from Center Front to the side seam. On paper, draw a length guideline and square a cross guideline from it. Place Center Front of the pattern on the length guideline, with the cross control line of the pattern on the cross guideline. Trace the Center Front—yoke seam intersection.

Then shift the pattern out ⅛" from the length guideline. Trace the yoke seam to just past the shirring notches. Mark the notches.

10   CROSS BUST GRADE*

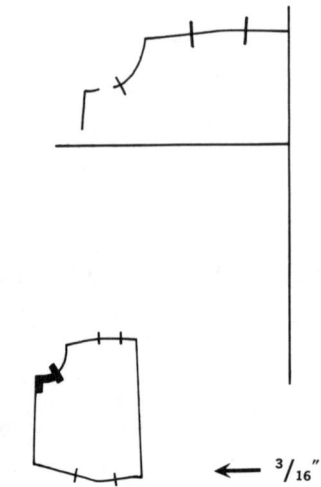

Shift the pattern out ³⁄₁₆" from its previous position. The pattern should now be a total of ⅜" from the length guideline. Trace the lower armhole and side seam intersection. Blend the armhole as shown on page 21. Mark the armhole notch.

1. In all steps except 4* and 10*, the measurements used are standard from one size to the next, no matter if it be a 1″, 1½″, or 2″ grade. Refer to the bodice grading chart for measurements for steps 4 and 10 *when doing other* than the 1½″ grade shown below, and follow the directions given for the Front Bodice.

2. The smaller diagram in the lower corner of each box depicts the pattern being graded, with the section to be traced in that step marked in heavy lines.

5 YOKE ARMHOLE GRADE

Shift the pattern down until the cross control line of the pattern lies on top of the cross guideline on the paper. The total downward shift of the pattern from its previous position will be approximately ¹⁄₁₆″. Trace the armhole to and around the yoke seam intersection.

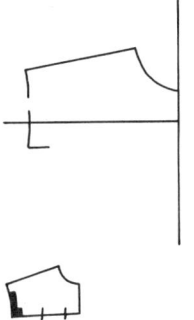

approx. ↓ ¹⁄₁₆″

6 YOKE SEAM GRADE

Shift the pattern in until it is within ⅛″ of the length guideline. Trace the yoke seam and mark the shirring notches. Blend the yoke seam across to the length guideline, thus establishing the new yoke seam—Center Front intersection.

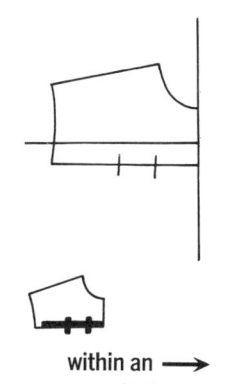

within an → ⅛″

8 YOKE SEAM GRADE

Shift the pattern out ¹⁄₁₆″ from its previous position. The pattern should now be a total of ³⁄₁₆″ from the length guideline. Trace the remainder of the yoke seam to and around the armhole intersection.

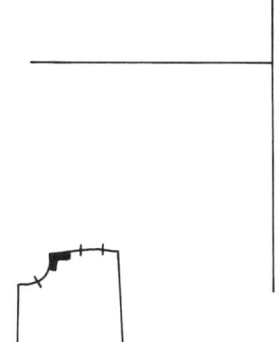

← ¹⁄₁₆″

9 ARMHOLE GRADE

Shift the pattern down so that the cross control line of the pattern is ⅛″ below the cross guideline on the paper. Trace the middle of the armhole.

↓ ⅛″

11 SIDE SEAM GRADE

Shift the pattern down ⅛″ from its previous position. The cross control line of the pattern should now be a total of ¼″ below the cross guideline on the paper. Trace the side seam to and around the waistline intersection.

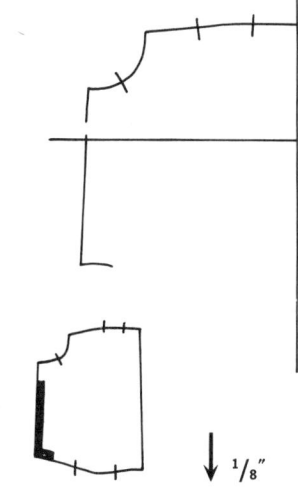

↓ ⅛″

12 APEX WIDTH GRADE

Shift the pattern in until it is within ⅛″ of the length guideline. Trace the waistline, and mark the shirring notches. Blend the waistline across to the length guideline, thus establishing the new Center Front—waistline intersection.

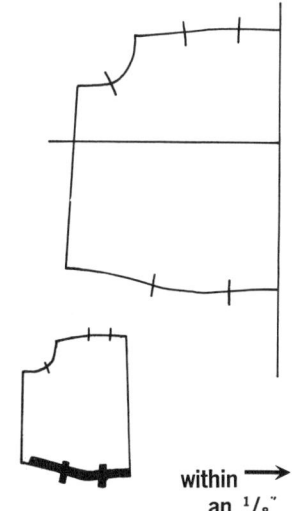

within → an ⅛″

HOW THE BODY GROWS

BODICE AND MIDRIFF

1½″ Grade

1. In order to maintain the original proportion of the design, the midriff section is not graded in length; the bodice portion of the pattern receives the entire length grade.

2. There is no increase given between the shirring notches, thus preserving the original fullness as created by the designer.

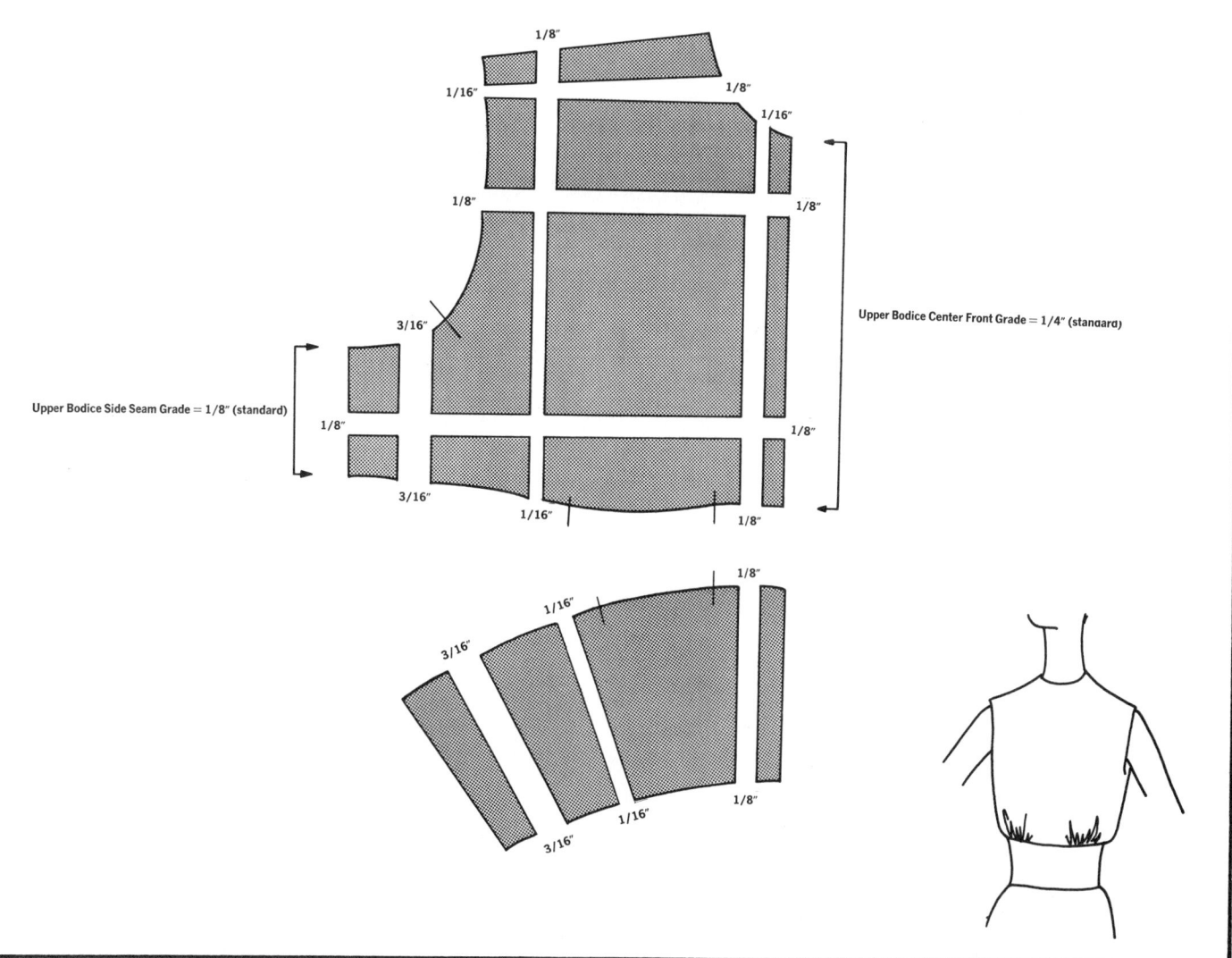

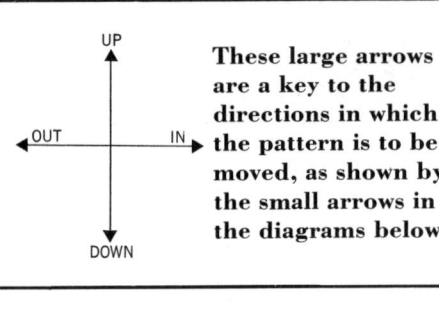

These large arrows are a key to the directions in which the pattern is to be moved, as shown by the small arrows in the diagrams below.

DIRECTIONS FOR A BODICE AND MIDRIFF 1½ INCH GRADE

1 to 6 BODICE GRADE

Draw a cross control line on the pattern by squaring a line from Center Front to the underarm-side seam intersection. On paper, draw a length guideline, and square a cross guideline from it.

Follow steps one through six of the front bodice grade on pages 26 and 27.

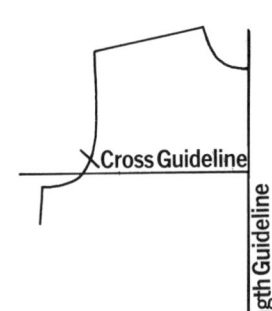
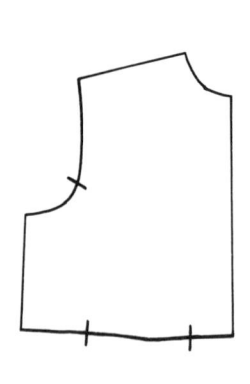
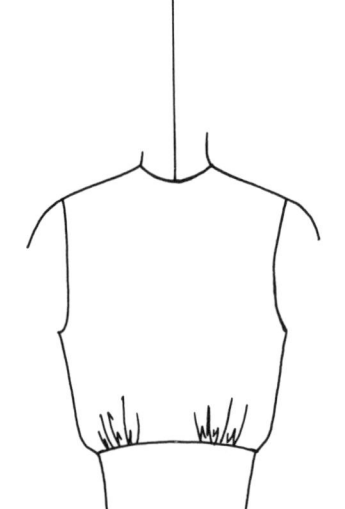

9 MIDRIFF GRADE

On paper, draw a length guideline and place Center Front of the midriff pattern on it. Trace the midriff seam-Center Front intersection. Then shift the pattern out ⅛″ from the length guideline. Trace the midriff seam to just past the midriff notches. Mark the notches.

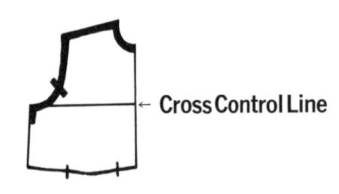

1. In all steps except 10*, the measurements used are standard from one size to the next no matter if it be a 1″, 1½″, or 2″ grade. Refer to column 6 of the bodice grading chart for the measurement for step 10 *when doing other than* the 1½″ grade shown below, and subtract ⅛″ from the measurement before using it.

2. The smaller diagram in the lower corner of each box depicts the pattern being graded, with the section to be traced in that step marked in heavy lines.

7 SIDE SEAM GRADE	Shift the bodice pattern down ⅛″ from its previous position. The control line of the pattern should now be a total of ¼″ below the cross guideline on the paper. Trace the side seam to and around the midriff seam intersection.

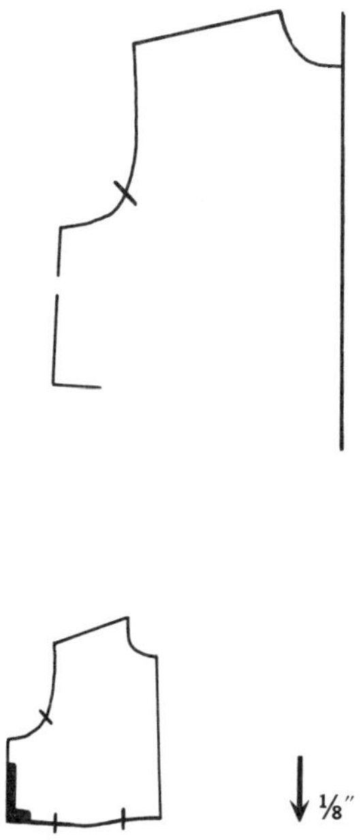

↓ ⅛″

8 MIDRIFF SEAM GRADE	Shift the bodice pattern in until it is within ⅛″ of the length guideline. Trace the midriff seam and mark the shirring notches. Blend the waistline across to the length guideline, thus establishing the new Center Front-midriff intersection. The back bodice is graded in a similar manner.

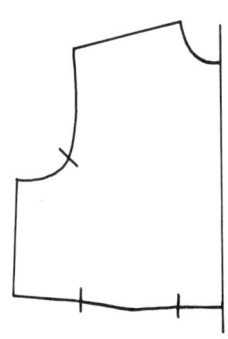

within ⅛″ →

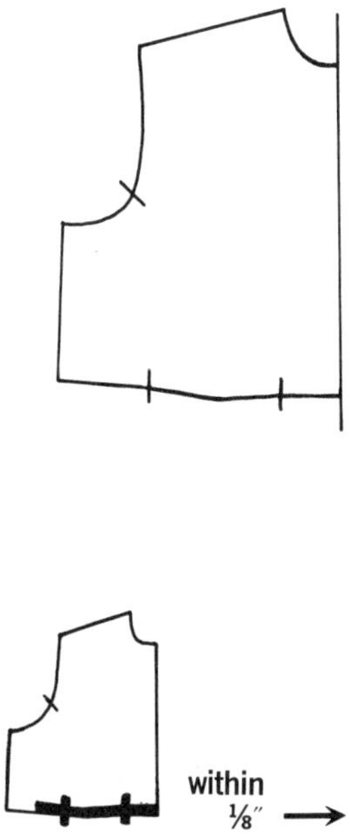

10 CROSS MIDRIFF GRADE*	Shift the midriff pattern out ¼″ from its previous position. The pattern should now be a total of ⅜″ from the length guideline. Trace the remainder of the midriff seam, the side seam, and the waistline intersection.

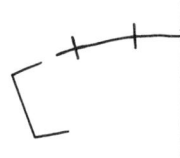

← ¼″

11 WAISTLINE GRADE	Shift the midriff pattern in until Center Front of the pattern meets the length guideline. On a 1½″ grade, the total shift of the pattern from its previous position will be ⅜″. Trace the waistline across to Center Front. The back midriff is graded in a similar manner to the front.

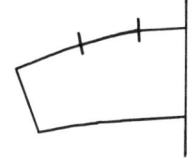

⅜″ →

HOW THE BODY GROWS

PRINCESS BODICE

1½" Grade

1. The princess bodice pattern is split at the front and back dart lines, and thus the total shoulder grade is divided equally between each shoulder section.

2. The princess sections are shifted up and down from the apex level, so that the corresponding princess seam grades can be distributed above and below the apex level.

3. The upper princess seam of the side section grades slightly longer than the corresponding seam of the center section. This extra length is used for ease on the side section.

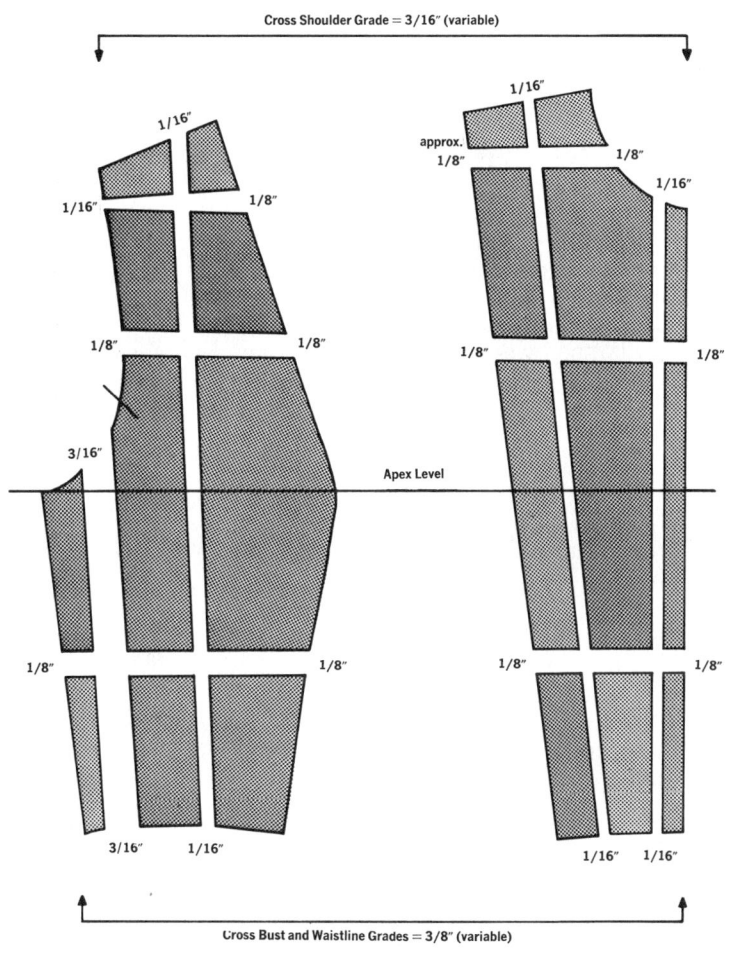

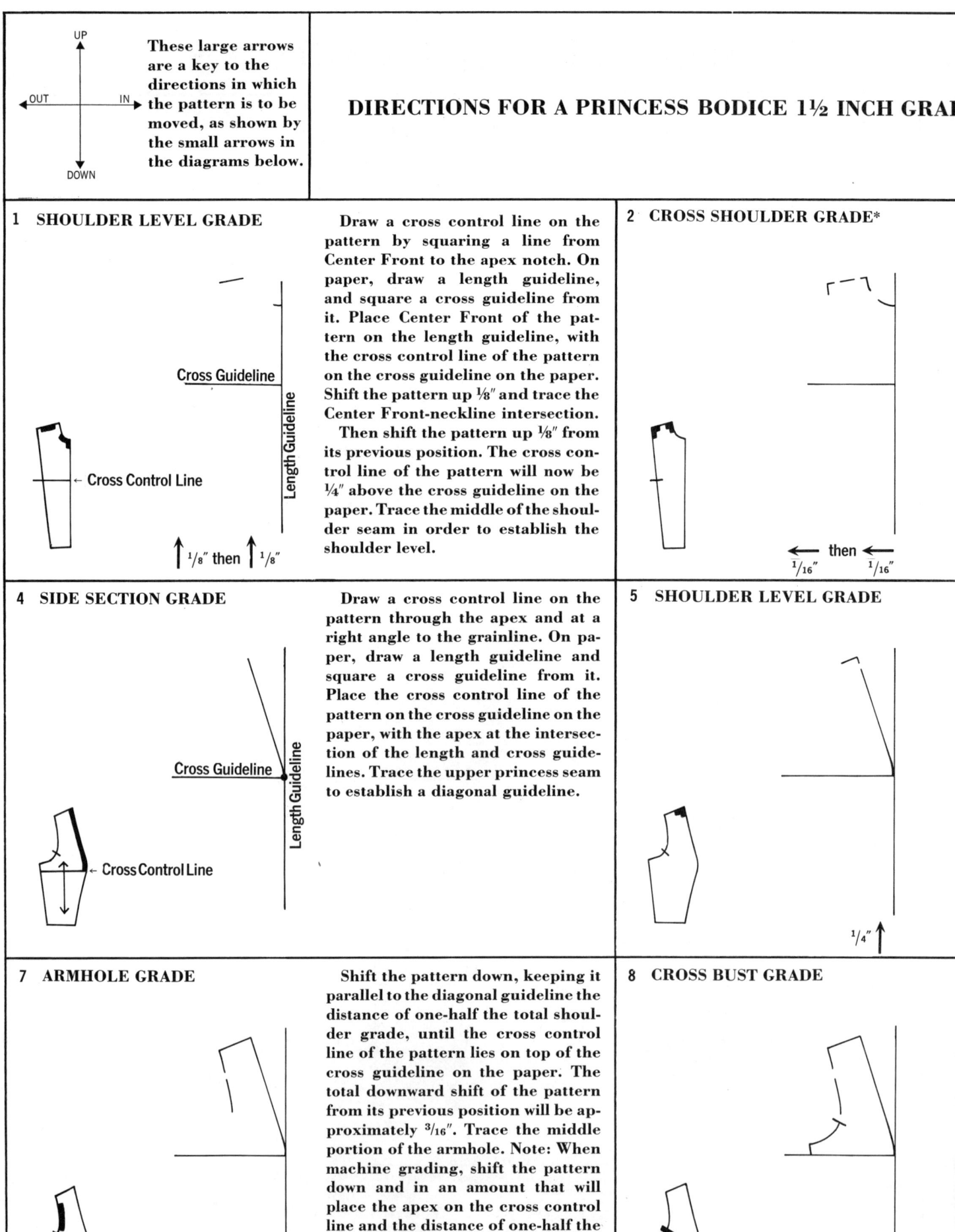

DIRECTIONS FOR A PRINCESS BODICE 1½ INCH GRADE

These large arrows are a key to the directions in which the pattern is to be moved, as shown by the small arrows in the diagrams below.

1 SHOULDER LEVEL GRADE

Draw a cross control line on the pattern by squaring a line from Center Front to the apex notch. On paper, draw a length guideline, and square a cross guideline from it. Place Center Front of the pattern on the length guideline, with the cross control line of the pattern on the cross guideline on the paper. Shift the pattern up ⅛" and trace the Center Front-neckline intersection.

Then shift the pattern up ⅛" from its previous position. The cross control line of the pattern will now be ¼" above the cross guideline on the paper. Trace the middle of the shoulder seam in order to establish the shoulder level.

↑ ⅛" then ↑ ⅛"

2 CROSS SHOULDER GRADE*

← 1/16" then ← 1/16"

4 SIDE SECTION GRADE

Draw a cross control line on the pattern through the apex and at a right angle to the grainline. On paper, draw a length guideline and square a cross guideline from it. Place the cross control line of the pattern on the cross guideline on the paper, with the apex at the intersection of the length and cross guidelines. Trace the upper princess seam to establish a diagonal guideline.

5 SHOULDER LEVEL GRADE

¼" ↑

7 ARMHOLE GRADE

Shift the pattern down, keeping it parallel to the diagonal guideline the distance of one-half the total shoulder grade, until the cross control line of the pattern lies on top of the cross guideline on the paper. The total downward shift of the pattern from its previous position will be approximately 3/16". Trace the middle portion of the armhole. Note: When machine grading, shift the pattern down and in an amount that will place the apex on the cross control line and the distance of one-half the total shoulder grade from the length guideline.

↓ approx. 3/16"

8 CROSS BUST GRADE

← 3/16"

1. In all steps except the second shift of step 2*, and step 6*, the measurements used are standard from one size to the next no matter if it be a 1″, 1½″, or 2″ grade. Refer to the Bodice grading chart for the measurements for the second shift of step 2, and step 6 *when doing other than the 1½″ grade shown below, and follow directions given for the front bodice.*

2. The smaller diagram in the lower corner of each box depicts the pattern being graded, with the section to be traced in that step marked in heavy lines.

Shift the pattern out ¹⁄₁₆″ from the length guideline, maintaining shoulder level. Trace the neckline-shoulder intersection, and blend the neckline as shown on page 20.

Then shift the pattern out ¹⁄₁₆″ (one-half the total shoulder grade) from its previous position, maintaining shoulder level. On a 1½″ grade, Center Front of the pattern will now be a total of ⅛″ from the length guideline. Trace the shoulder seam to and around the upper princess seam intersection.

3 PRINCESS SEAM GRADE

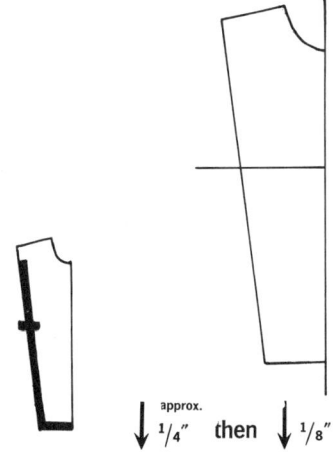

approx. ↓ ¼″ then ↓ ⅛″

Shift the pattern down until the cross control line lies on the cross guideline. The total shift will be approximately ¼″ from its previous position. Blend a line from the end of the shoulder to the apex notch. Crossmark the apex notch.

Then shift the pattern down so that the cross control line is ⅛″ below the cross guideline. Trace the princess seam-waistline intersection, and blend the princess seam from the waistline to the shoulder. Blend the waistline across to the length guideline, thus establishing the new Center Front-waistline intersection.

NOTE: On a 1½″ grade, the pattern will automatically be within ⅛″ of Center Front. On a 1″ or 2″ grade, adjust the pattern so that it is within ⅛″ of Center Front, before tracing the princess seam-waistline intersection.

Shift the pattern up *along the diagonal* ¼″, maintaining, the cross control line of the pattern parallel to the cross guideline on the paper. Trace the upper princess seam to and around the shoulder to the middle of the shoulder seam, thus establishing the shoulder level. Note: When machine grading, mark ¼″, by ruler, up the diagonal from the apex, and then shift the pattern out and up an amount that will place the apex on the ¼″ mark.

6 CROSS SHOULDER GRADE*

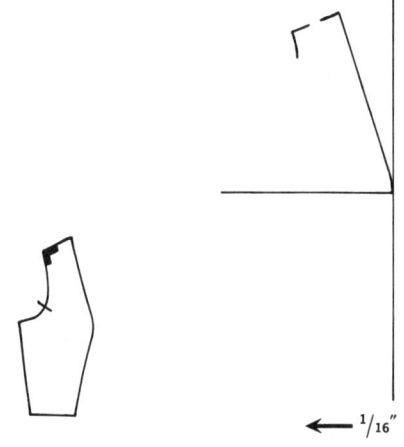

← ¹⁄₁₆″

Shift the pattern out ¹⁄₁₆″ (one-half the total shoulder grade) from its previous position, maintaining shoulder level. The upper princess seam must remain parallel to the traced diagonal guideline. Trace the remainder of the shoulder seam to and around the armhole intersection.

Shift the pattern out ³⁄₁₆″ from its previous position. On a 1½″ grade, the apex of the pattern should now be a total of ¼″ from the length guideline. Trace the lower armhole and the side seam intersection. Blend the armhole as shown on page 21. Mark the armhole notch. For a 1″ grade, shift out ³⁄₃₂″. For a 2″ grade, shift out ⁹⁄₃₂″.

9 SIDE SEAM AND WAISTLINE GRADE

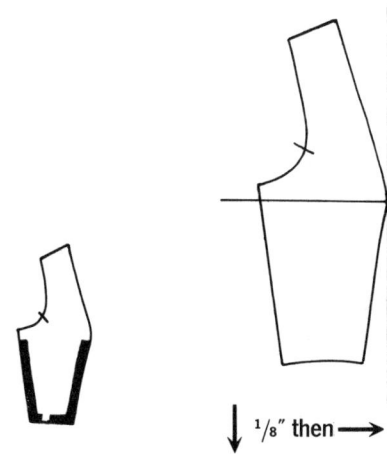

↓ ⅛″ then →

Shift the pattern down until the cross control line of the pattern is ⅛″ below the cross guideline on the paper. Trace the side seam to and around the waistline intersection.

Then shift the pattern in until the apex of the pattern touches the length guideline. Trace the waistline to and around the lower princess seam, and blend to the apex. The side bodice grade is now complete.

The back princess bodice is graded in a similar manner.

HOW THE BODY GROWS

SIX-GORE SKIRT

1½″ Grade

1. The six-gore skirt is graded the same amount as the basic two-piece skirt even though it is cut into gores.

2. The waistline of the Center Front and Center Back panels grade ⅛″ per size to assure alignment with the bodice darts, when the skirt is part of an entire garment. The balance of the total waistline grade is given to the side gore.

3. The lengthwise increase of ¼″ per size may be held after Misses size 14 and Junior size 13.

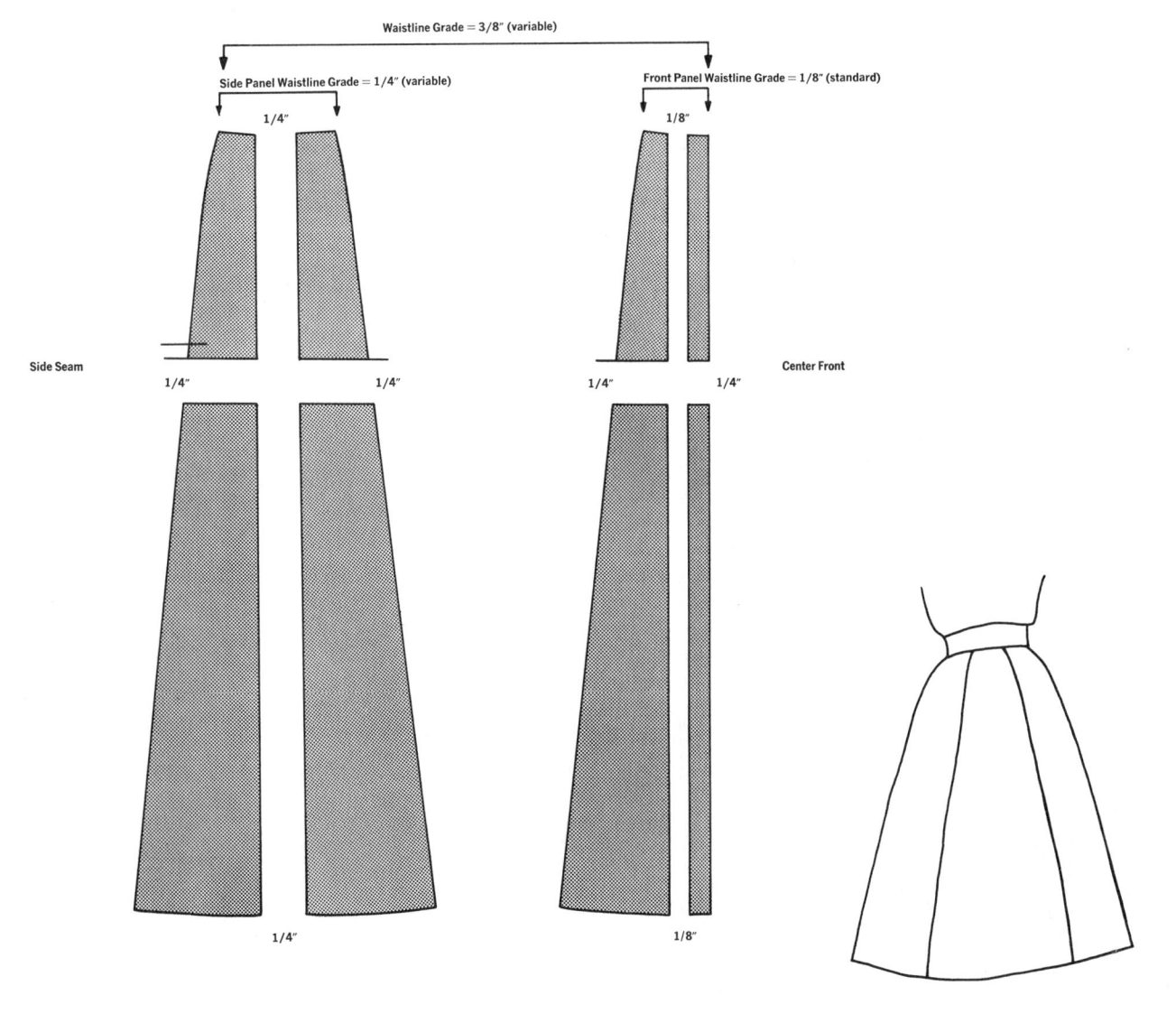

These large arrows are a key to the directions in which the pattern is to be moved, as shown by the small arrows in the diagrams below.

DIRECTIONS FOR A SIX-GORED SKIRT 1½ INCH GRADE

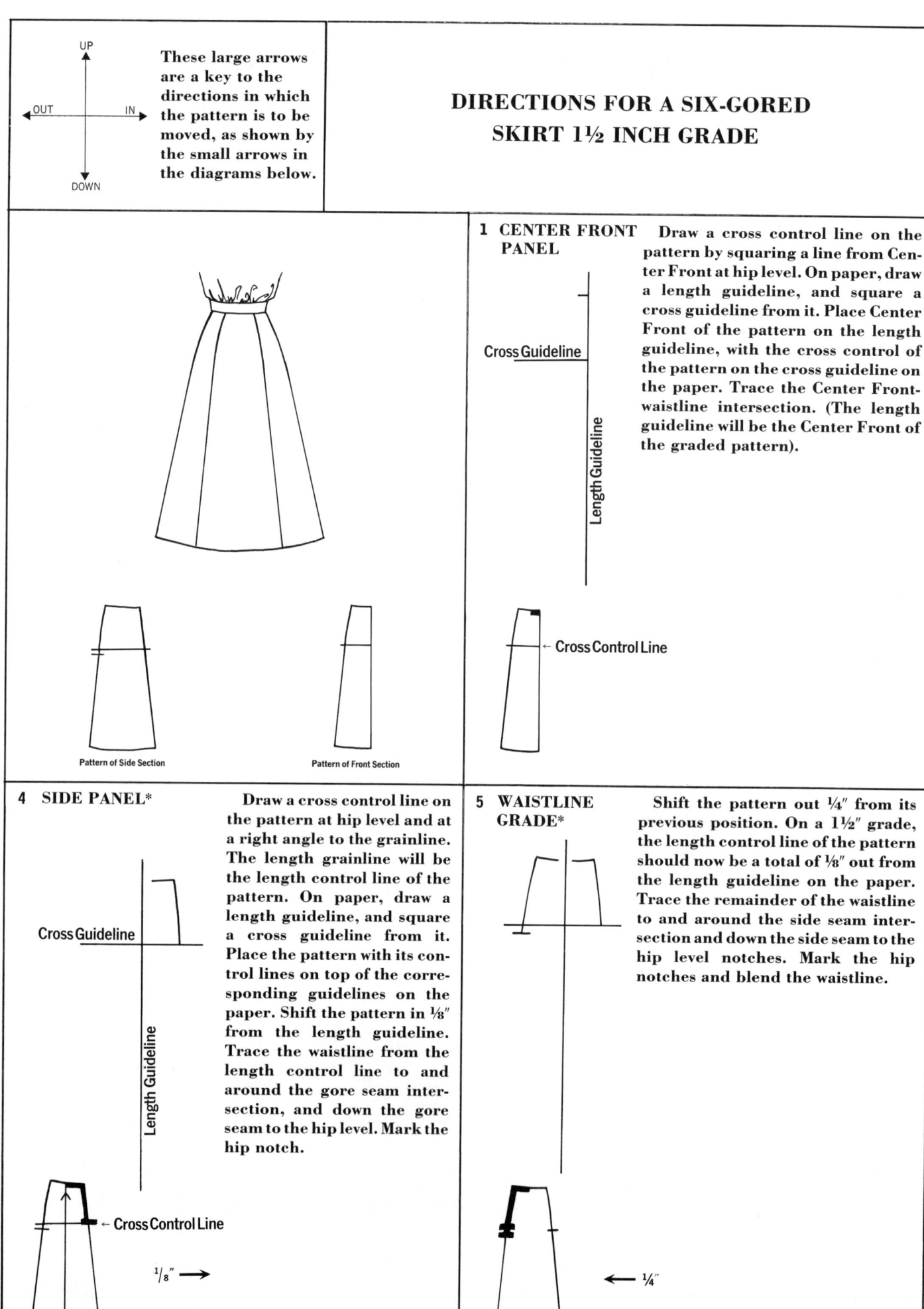

1 CENTER FRONT PANEL Draw a cross control line on the pattern by squaring a line from Center Front at hip level. On paper, draw a length guideline, and square a cross guideline from it. Place Center Front of the pattern on the length guideline, with the cross control of the pattern on the cross guideline on the paper. Trace the Center Front-waistline intersection. (The length guideline will be the Center Front of the graded pattern).

4 SIDE PANEL* Draw a cross control line on the pattern at hip level and at a right angle to the grainline. The length grainline will be the length control line of the pattern. On paper, draw a length guideline, and square a cross guideline from it. Place the pattern with its control lines on top of the corresponding guidelines on the paper. Shift the pattern in ⅛" from the length guideline. Trace the waistline from the length control line to and around the gore seam intersection, and down the gore seam to the hip level. Mark the hip notch.

5 WAISTLINE GRADE* Shift the pattern out ¼" from its previous position. On a 1½" grade, the length control line of the pattern should now be a total of ⅛" out from the length guideline on the paper. Trace the remainder of the waistline to and around the side seam intersection and down the side seam to the hip level notches. Mark the hip notches and blend the waistline.

1. In all steps except 4*, 5* and 7*, the measurements used are standard from one size to the next no matter if it be a 1″, 1½″, or 2″ grade. Refer to column 7 of the skirt grading chart for the measurements for steps 4 and 5 *when doing other* than the 1½″ grade shown below. For step 4, subtract ⅛″ from the measurement, and divide the remainder in half before using it. For steps 5 and 7, subtract ⅛″ from the measurement before using it.

2. The smaller diagram in the lower corner of each box depicts the pattern being graded, with the section to be traced in that step marked in heavy lines.

2 WAISTLINE GRADE

Shift the pattern out ⅛″ from the length guideline. Trace the waistline to and around the gore seam intersection and down the gore seam to the hip level. Mark the hip notch.

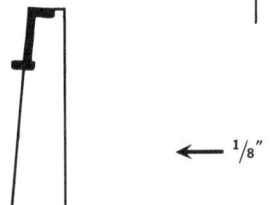
← ⅛″

3 GORE SEAM GRADE

Shift the pattern down until the cross control line of the pattern is ¼″ below the cross guideline on the paper. Trace the remainder of the gore seam to and around the hemline intersection, and blend the hemline across to the length guideline.

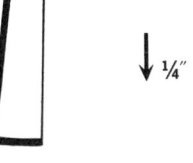
↓ ¼″

6 SIDE SEAM GRADE

Shift the pattern down until the cross control line of the pattern is ¼″ below the cross guideline of the paper. Trace the side seam to and around the hemline intersection.

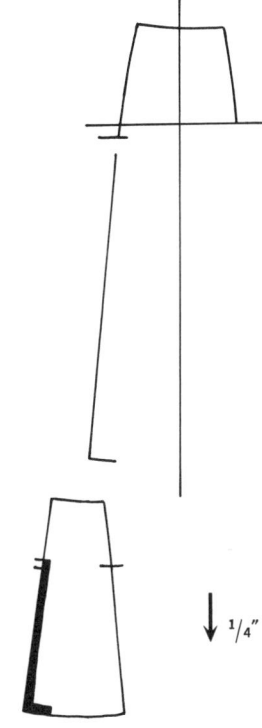
↓ ¼″

7 GORE SEAM GRADE*

Shift the pattern in ¼″ from its previous position. The length control line of the pattern should now be a total of ⅛″ in from the length guideline on the paper. Trace the hemline to and around the gore seam intersection, and blend the gore seam up to the hip level.

The back skirt gores are graded in a similar manner.

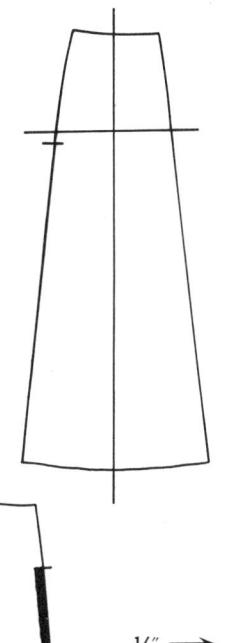
¼″ →

HOW THE BODY GROWS

CIRCLE SKIRT | 1½" Grade

1. The circle skirt receives the same waistline grade as a straight skirt of the same size, distributed proportionally along the waistline as shown below in the diagram.

2. The side seam and Center Front lengths may be held after size 14.

3. The increased hem flare per size is the sum of the side seam, Center Front, and waistline increases.

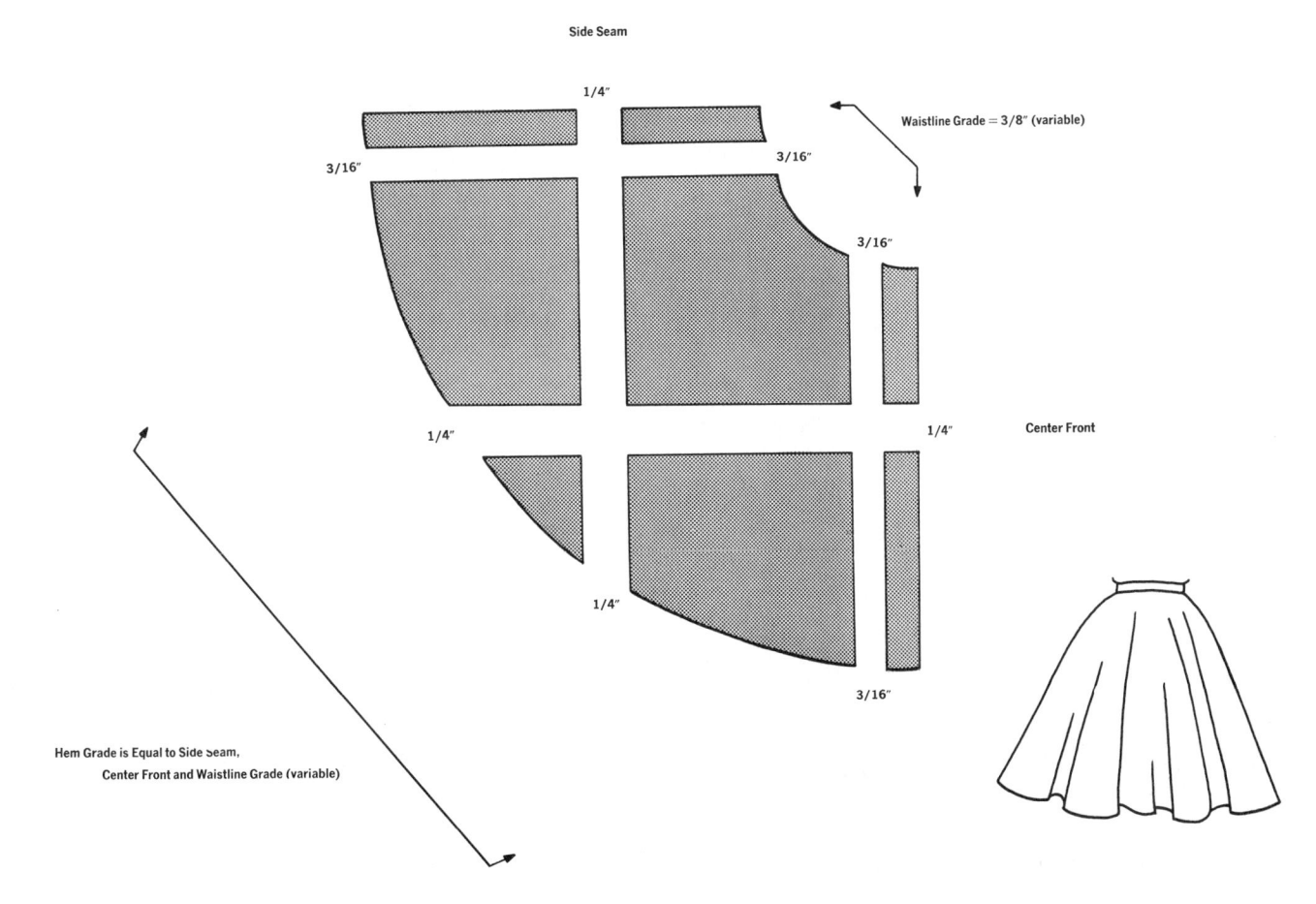

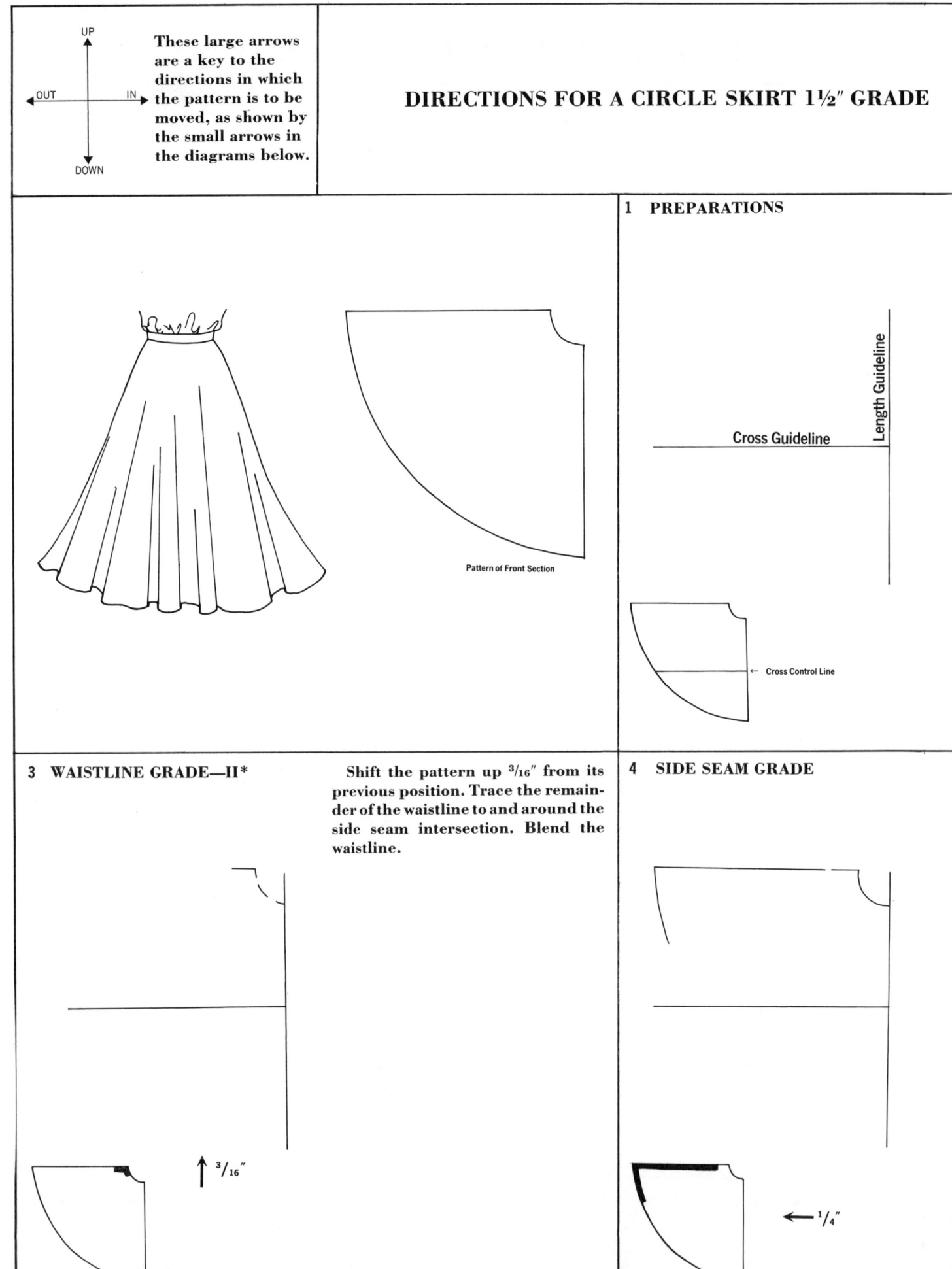

These large arrows are a key to the directions in which the pattern is to be moved, as shown by the small arrows in the diagrams below.

DIRECTIONS FOR A CIRCLE SKIRT 1½" GRADE

1 PREPARATIONS

Pattern of Front Section

Length Guideline

Cross Guideline

← Cross Control Line

3 WAISTLINE GRADE—II*

Shift the pattern up ³/₁₆" from its previous position. Trace the remainder of the waistline to and around the side seam intersection. Blend the waistline.

↑ ³/₁₆"

4 SIDE SEAM GRADE

← ¼"

1. In all steps except 2* and 3*, the measurements used are standard from one size to the next, no matter if it be a 1", 1½", or 2" grade. For steps 2 and 3, shift out one half of the waistline increase.

2. The smaller diagram in the lower corner of each box depicts the pattern being graded, with the section to be traced in that step marked in heavy lines.

Draw a cross control line on the pattern by squaring a line from Center Front. On the paper, draw a length guideline and square a cross guideline from it.

2 WAISTLINE GRADE—I*

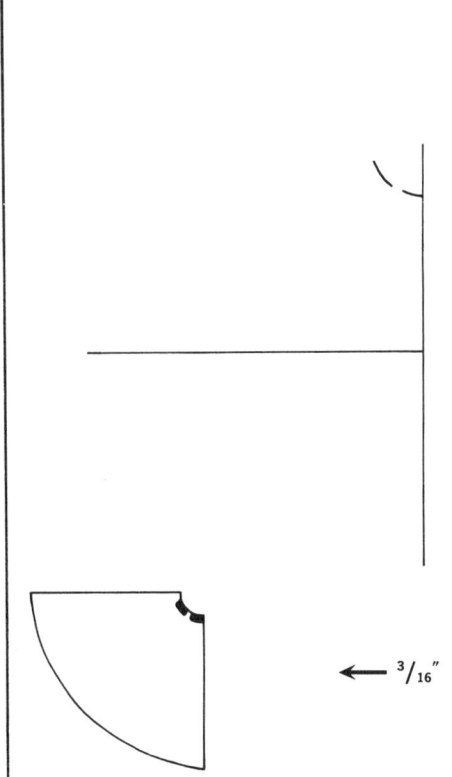

← ³⁄₁₆"

Place Center Front of the pattern on the length guideline, with the cross control line of the pattern on the cross guideline. Trace the Center Front—waistline intersection. (The length guideline will be the Center Front of the graded pattern).

Then shift the pattern out ³⁄₁₆" from the length guideline. Trace the middle of the waistline seam.

Shift the pattern out ¼" from its previous position. On a 1½" grade, the pattern should now be a total of ⁷⁄₁₆" from the length guideline. Trace the side seam and approximately one-fourth of the skirt hemline. (If the skirt is to be held in length, omit this outward shift, and do just the indicated tracing.)

5 CENTER FRONT GRADE

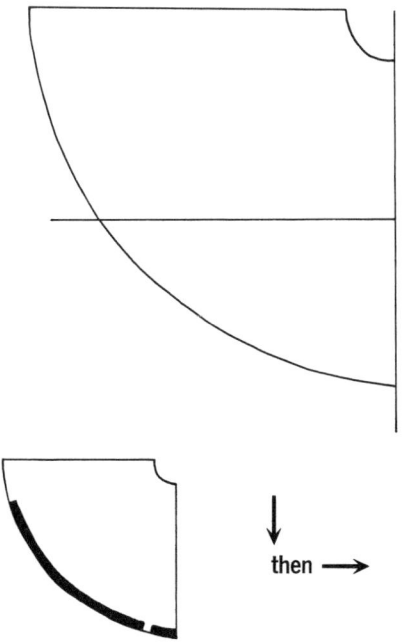

↓
then →

Shift the pattern down so that the cross control line of the pattern is ¼" below the cross guideline on the paper. Trace nothing. (If the skirt is to be held in length, shift the pattern down until the cross control line lies on top of the cross guideline).

Then shift the pattern in until the Center Front of the pattern lies on top of the length guideline on the paper. Trace the Center Front—hemline intersection and blend the hemline. The back circle skirt is graded in a similar manner.

THE FOUR GORE SKIRT

The four gore skirt is graded in a similar manner to a straight skirt as shown on pages 38 and 39, even though there may be no darts in the waistline. It is best to maintain the contour of the waistline by evenly distributing the grade along the entire waistline curve.

THE EIGHT GORE SKIRT

The eight gore skirt is graded in a similar manner to the six gore skirt, as shown on pages 68 and 69, except that the entire waistline grade is divided by 8 to determine the grade for each gore. For example: A 1½" waistline grade divided by 8 equals a ³⁄₁₆" grade per gore. Follow the directions given for the *side* gore of the six gore skirt for *each* pattern piece of the eight gore skirt, (as they are all similarly shaped), substituting one-half of the new measurement in step 4, and the total new measurement in step 5 of the gore skirt lesson.

GRADING SKIRT VARIATIONS

THE PEG SKIRT

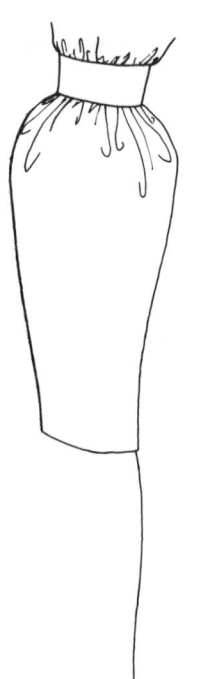

THE DIRNDL SKIRT

The dirndl skirt is usually not graded in width from size to size, due to the gathered waistline which can, up to a point, be adjusted to larger sizes without grading. For all skirts, it is usually the practice to hold the skirt length after Misses size 14 and Junior size 13.

The peg skirt is graded in a similar manner to a straight skirt, distributing the grade along the waistline in order to maintain the contour of the curve of the waistline seam. For all skirts, the waistband is usually not graded in width, but is graded exactly the same as the waistline in length.

GORE PLEATED SKIRT

The six gore pleated skirt is graded in the same manner as the six gore skirt. The pleats remain the same from size to size, since their depth as planned by the designer has no relationship to the size of the skirt itself.

GATHERED FRONT SIX GORE SKIRT

This gathered front skirt is graded in the same manner as the six gore skirt, as are all variations of the six gore skirt, such as the side gathered or side flared skirt.

GRADING SKIRT VARIATIONS

YOKE SKIRT

The hip yoke skirt is graded as two skirts. The yoke section is graded in the same manner as the straight skirt. The skirt section is graded the same as whichever skirt it is based upon, as in this case the six gore skirt. If the skirt is to be graded in length, the length grade should be given in the skirt section.

TRUMPET SKIRT

This trumpet skirt is graded in the same manner as described for the eight gore skirt. The amount of flare in a skirt does not change in any way the basic grading techniques that would be used on that style skirt without flares.

HOW THE BODY GROWS

SHAWL COLLAR | 1½″ Grade

1. The length of Center Back of the collar is not increased from one size to the next. The back neck increases of the collar are based on the neck grade of the back bodice.

2. A double breasted extension is graded in a similar manner to the single breasted closing.

3. A dart may be placed under the collar along the front fold line, if desired, after the grading is completed.

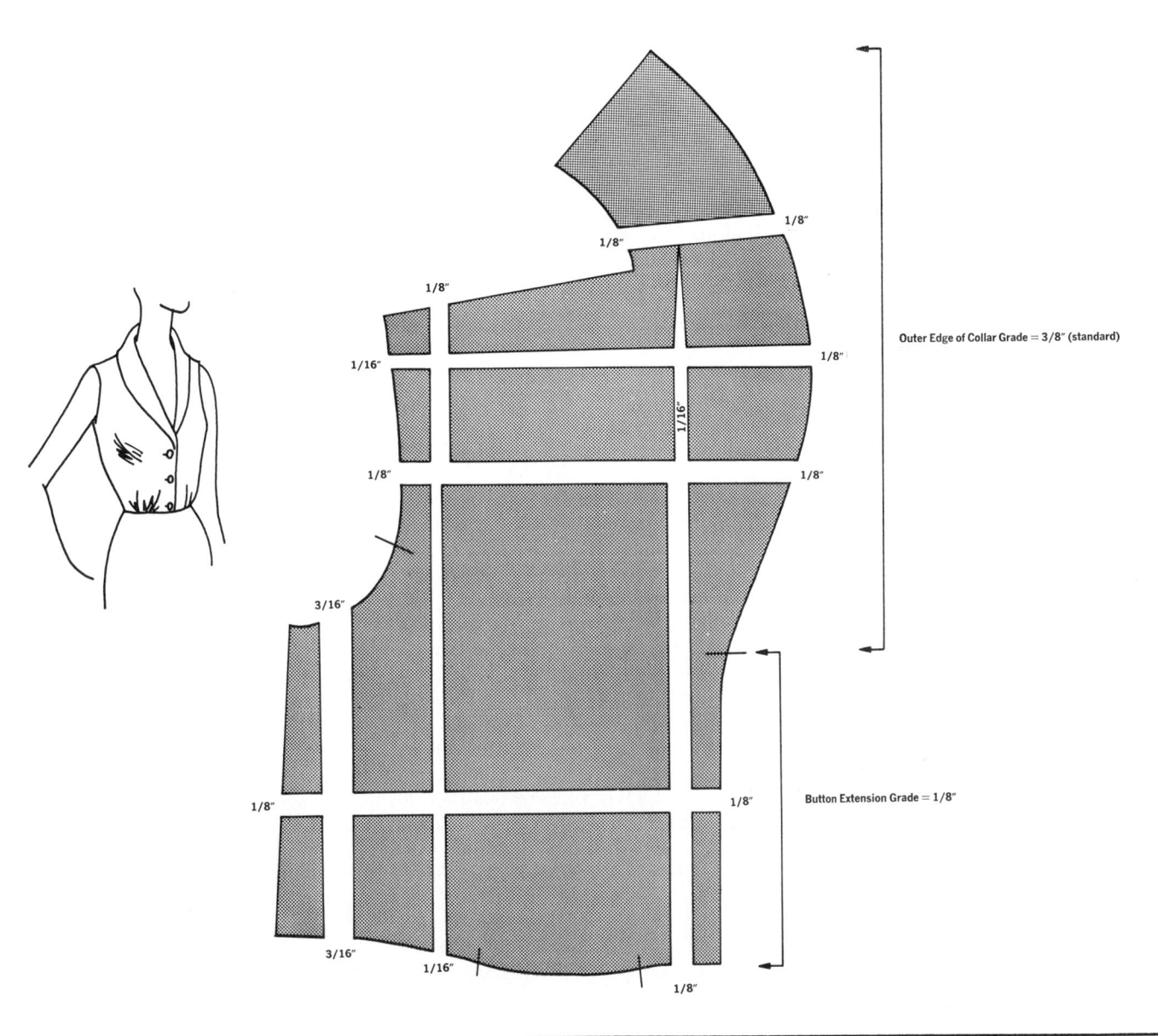

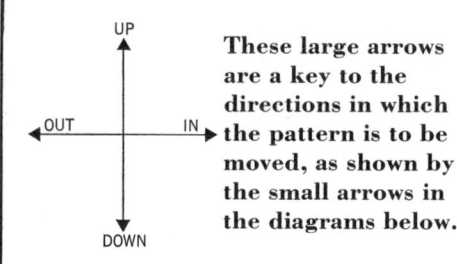

These large arrows are a key to the directions in which the pattern is to be moved, as shown by the small arrows in the diagrams below.

DIRECTIONS FOR A SHAWL COLLAR 1½″ GRADE

1 COLLAR PREPARATIONS

Mark a cross control line on the pattern by squaring a line from Center Front across to the side seam and to the button extension. Center Front will be the length control line of the pattern. On paper, draw a length guideline and square a cross guideline from it. Place Center Front of the pattern on the length guideline, with the cross control line of the pattern on the cross guideline on the paper. Trace the button extension from the waistline to just past the breakpoint crossmark. Mark the crossmark and the top button level.

2 FRONT LENGTH GRADE

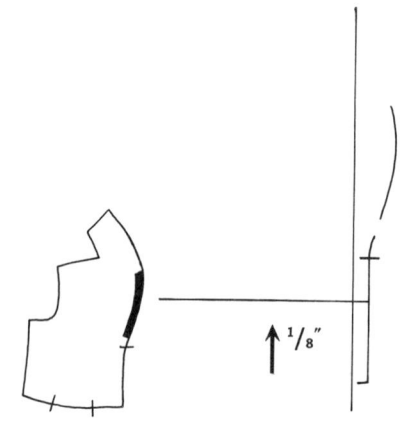

4 COLLAR GRADE—II

Shift the pattern up ⅛″ from its previous position. The cross control line of the pattern should now be a total of ⅜″ above the cross guideline on the paper. Trace the outer edge of the collar to and around Center Back, to the neckline intersection. Blend the outer edge of the collar.

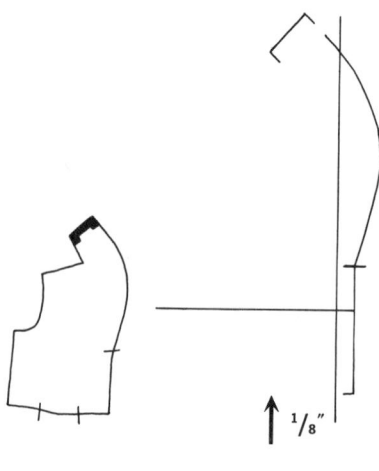

5 BACK NECKLINE GRADE—I

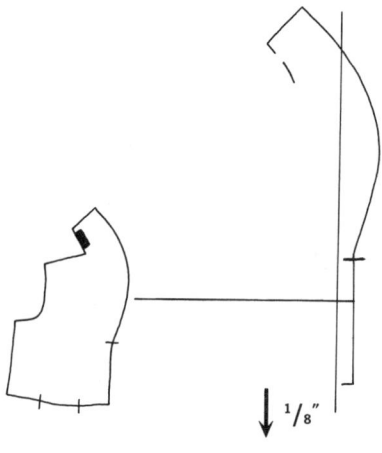

7 CROSS SHOULDER GRADE*

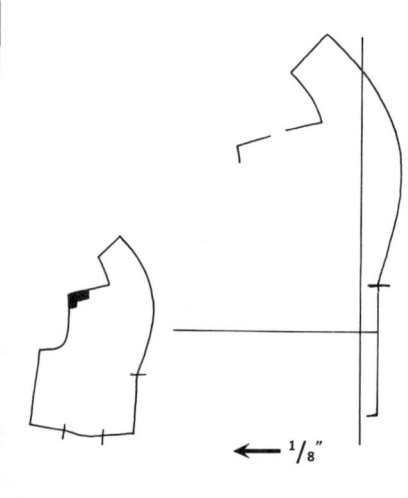

Shift the pattern out ⅛″ from its previous position, maintaining shoulder level. (Note: the maintaining of shoulder level in the cross shoulder step of the shawl bodice is used only for a 1½″ grade. Therefore, when doing a cross shoulder step on a 1″ or 2″ grade, the cross control line of the pattern must be placed ³⁄₁₆″ above the cross guideline on the paper, disregarding shoulder level, before doing the indicated tracing). On a 1½″ grade, the pattern should now be a total of ³⁄₁₆″ from the length guideline. Trace the remainder of the shoulder, and the armhole intersection. Blend the shoulder seam as shown on page 19.

8 ARMHOLE AND CROSSBUST GRADE*

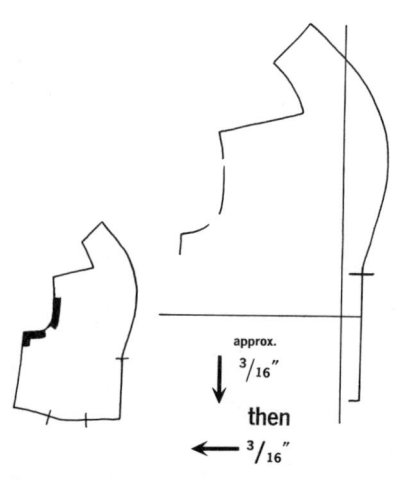

1. In all steps except 7* and the second shift of step 8*, the measurements used are standard from one size to the next, no matter if it be a 1″, 1½″, or 2″ grade. Refer to the bodice grading chart for the measurements for step 7 and the second shift of step 8 *when doing other* than the 1½″ grade shown below. For step 7 subtract column 3 from column 4 before applying the measurement. For the second shift of step 8, divide the measurement in column 6 in half before using.

2. The smaller diagram in the lower corner of each box depicts the pattern being graded, with the section to be traced in that step marked in heavy lines.

Shift the pattern up ⅛″ from its previous position. Trace the outer edge of the collar, ending below the shoulder level of the bodice.

3 COLLAR GRADE—I

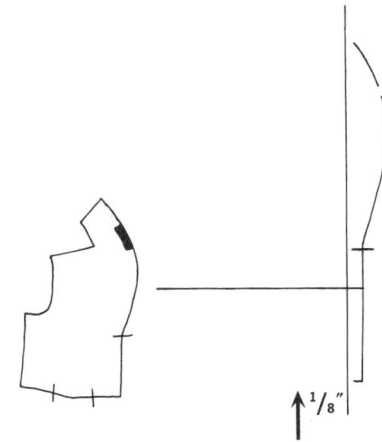

↑ ⅛″

Shift the pattern up ⅛″ from its previous position. The cross control line of the pattern should now be a total of ¼″ above the cross guideline on the paper. Trace the outer edge of the collar for a short distance, ending above the shoulder level of the bodice.

Shift the pattern down ⅛″ from its previous position. The cross control line of the pattern should now be a total of ¼″ above the cross guideline on the paper. Trace a short distance of the back neckline to a point just short of the shoulder intersection.

6 BACK NECKLINE GRADE—II

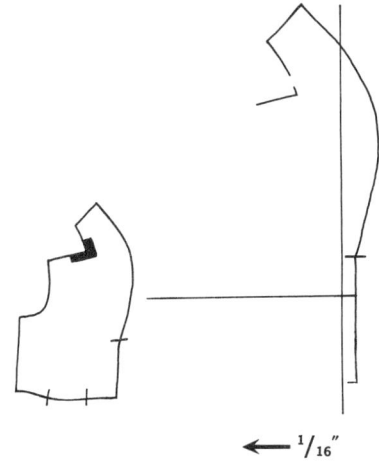

← 1/16″

Shift the pattern out 1/16″ from the length guideline. Trace the remainder of the back neckline to and around the shoulder intersection, and trace approximately one half of the shoulder seam, in order to establish the shoulder level. Blend the back neckline.

Shift the pattern down so that the cross control line of the pattern lies on top of the cross guideline on the paper. The total downward shift of the pattern from its previous position will be approximately 3/16″. Trace the middle of the armhole.

Then shift the pattern out 3/16″ from its previous position. On a 1½″ grade, the pattern should now be a total of ⅜″ from the length guideline. Trace the lower armhole and side seam intersection. Blend the armhole as shown on page 21. Mark the armhole notch.

9 SIDE SEAM AND APEX WIDTH GRADE

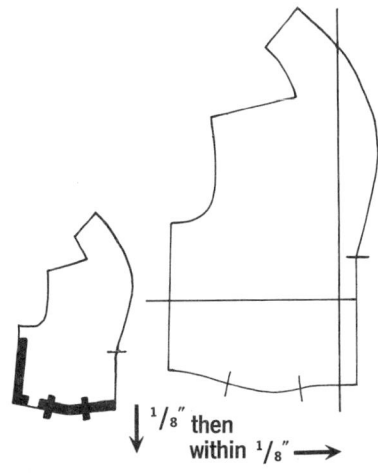

↓ ⅛″ then within ⅛″ →

Shift the pattern down ⅛″ from its previous position. The cross control line of the pattern should now be a total of ⅛″ below the cross guideline on the paper. Trace the side seam to and around the waistline intersection.

Then shift the pattern in until it is within ⅛″ of the length guideline. Trace the waistline, and mark the shirring notches and the bottom button level. Blend the waistline across to the length guideline, thus establishing the new Center Front—waistline intersection. Distribute the remaining button holes evenly between the top and bottom button levels. The shawl collar 1½″ grade is now complete.

PART 4

Grading Advanced Designs

KIMONO SLEEVE

KIMONO RAGLAN SLEEVE

SQUARE ARMHOLE

GUSSET SLEEVE

SET-IN RAGLAN SLEEVE

**PRINCESS BODICE
& SLEEVE IN ONE**

**DROPPED SHOULDER
SLEEVE**

HOW THE BODY GROWS

KIMONO SLEEVE

1½″ Grade

1. The usual side seam grade of ⅛″ per size is here graded *above* the actual side seam, as the grade is needed to achieve the proper width at the elbow. For this reason, the entire elbow grade is a standard ½″ per size on the kimono and gusset sleeves, rather than a variable grade as on the set-in sleeve.

2. The split diagram shows a front wrist-length fitted sleeve, with a straight line indicating the elbow. On a fitted sleeve, the wrist is always grade one-half of the elbow grade. For example, the split diagram shows the front elbow increasing ¼″; thus the wrist increases ⅛″. For a straight sleeve the wrist is graded the same amount as the elbow.

3. The waistline dart grades ⅛″ per size regardless of the perimeter changes of the bodice.

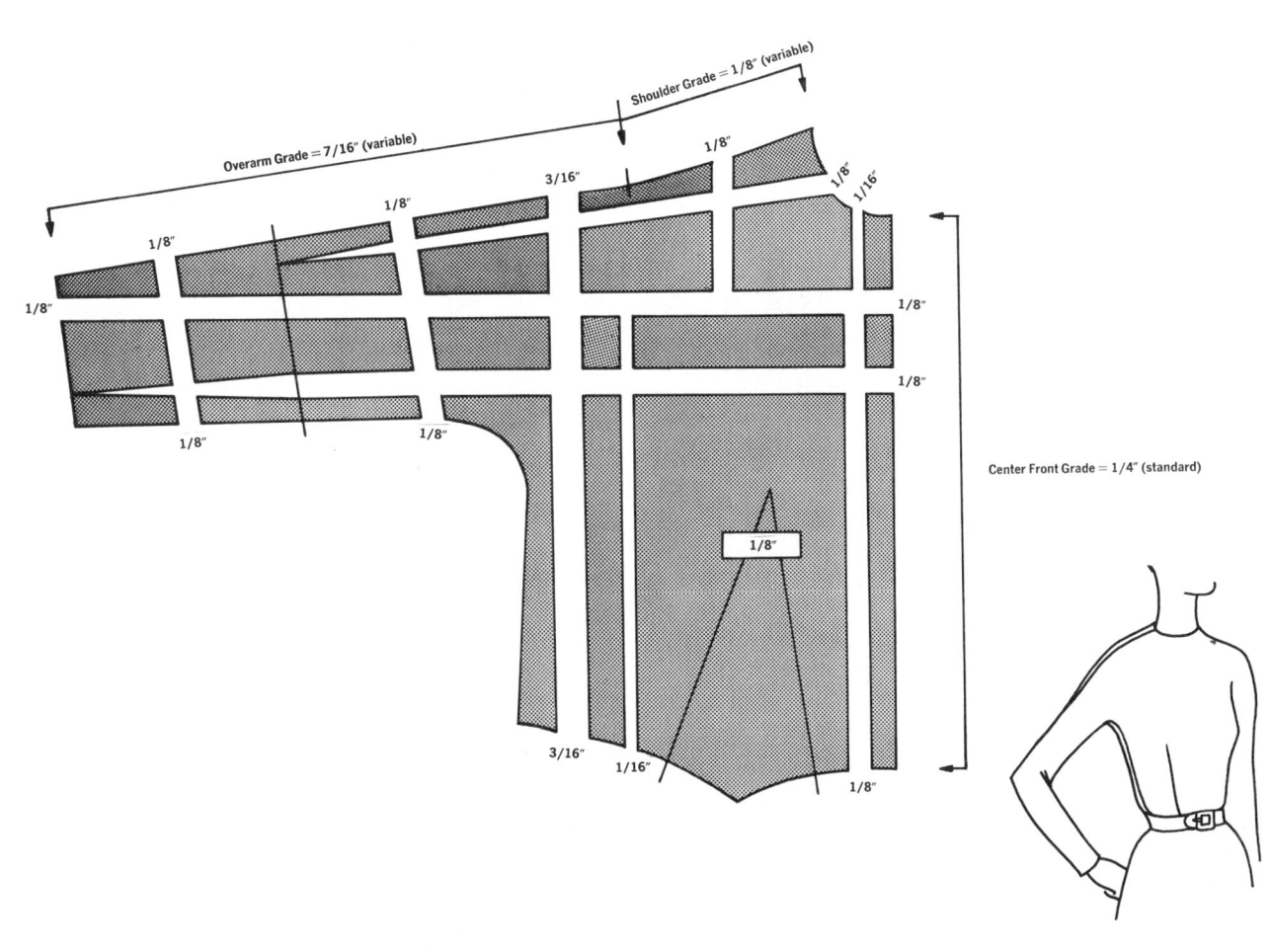

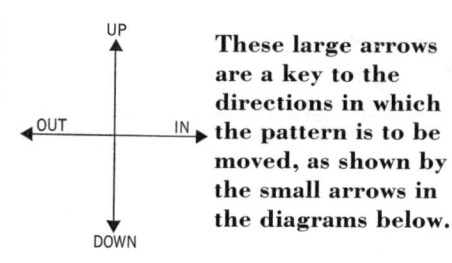

These large arrows are a key to the directions in which the pattern is to be moved, as shown by the small arrows in the diagrams below.

DIRECTIONS FOR A KIMONO SLEEVE 1½ INCH GRADE

1 SHOULDER LEVEL GRADE

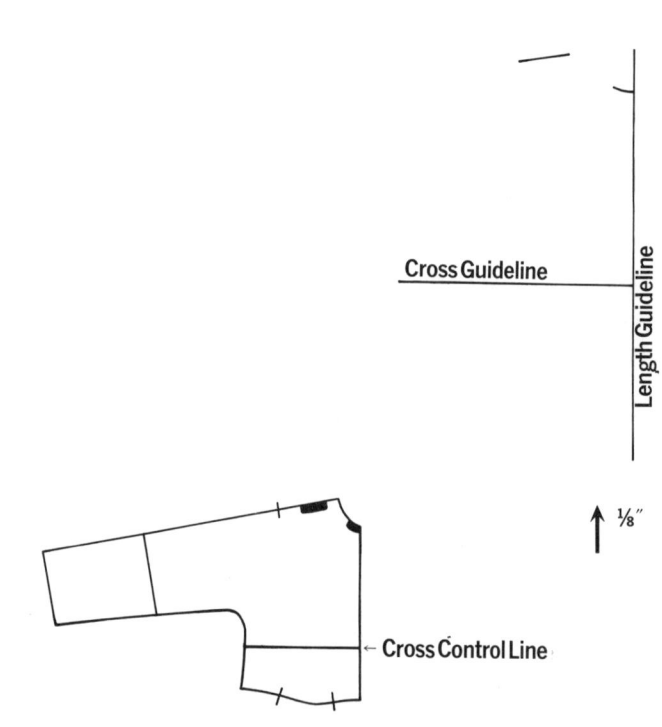

Draw a cross control line on the pattern by squaring a line from Center Front to the side seam. On paper, draw a length guideline, and square a cross guideline from it. Place Center Front of the pattern on the length guideline, with the cross control line of the pattern on the cross guideline on the paper. Trace the neckline-Center Front intersection. (The length guideline will be the Center Front of the graded pattern).

Shift the pattern up ⅛" on the length guideline. Trace the middle of the shoulder seam in order to establish the shoulder level.

3 CROSS SHOULDER GRADE*

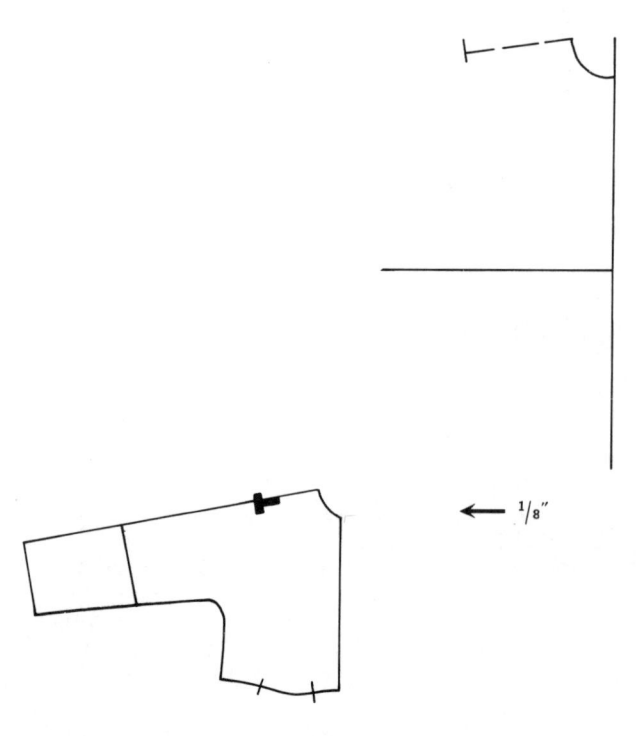

Shift the pattern out ⅛" from its previous position, maintaining shoulder level. On a 1½" grade, the pattern should now be a total of ³⁄₁₆" from the length guideline. Trace the remainder of the shoulder to the shoulder notch. Mark the shoulder notch.

1. In all steps except 3* and the second shift in step 4*, the measurements used are standard from one size to the next no matter if it be a 1″, 1½″, or 2″ grade. Refer to the bodice grading chart for the measurements for step 3 and the second shift in step 4 *when doing other* than the 1½″ grade shown below. For step 3 subtract column 3 from column 4 before applying the measurement. For the second shift in step 4, divide the measurement in half and add ⅛″ before using.

2. The smaller diagram in the lower corner of each box depicts the pattern being graded, with the section to be traced in that step marked in heavy lines.

2 NECK GRADE

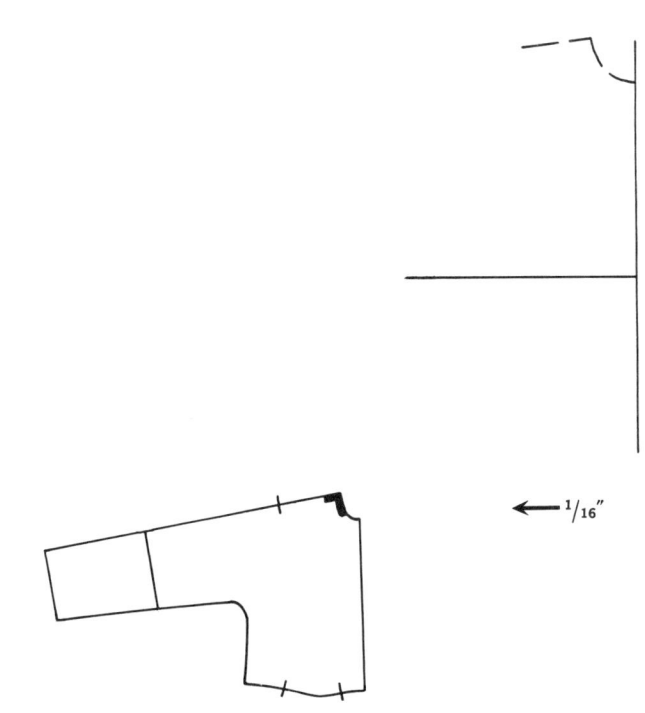

Shift the pattern out ¹⁄₁₆″ from the length guideline, maintaining shoulder level. Trace the shoulder-neckline intersection. Blend the neckline as shown on page 20.

4 CROSS BUST AND OVERARM GRADE*

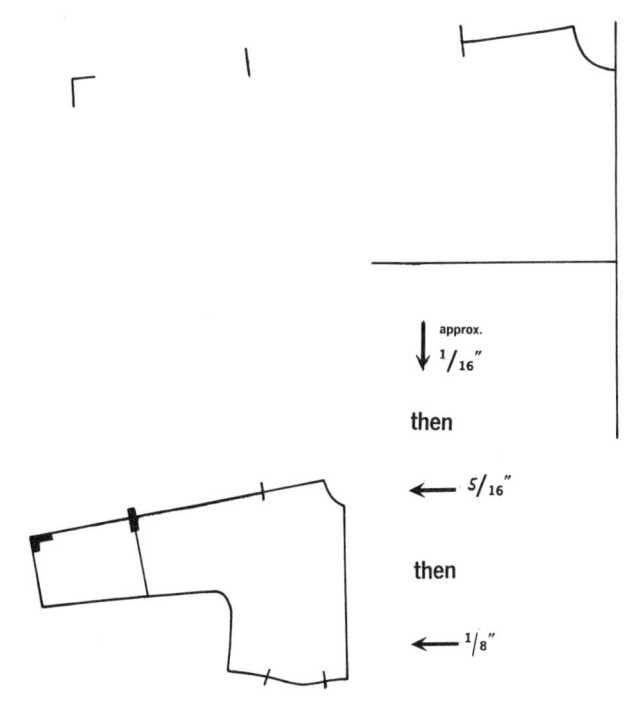

Shift the pattern down until the cross control line of the pattern lies on top of the cross guideline on the paper. The total downward shift of the pattern from its previous position will be approximately ¹⁄₁₆″.

Then shift the pattern out ⁵⁄₁₆″ from its previous position, maintaining cross guideline level. On a 1½″ grade, the pattern should now be a total of ½″ from the length guideline. Mark the elbow notch.

Then shift the pattern out ⅛″ from its previous position, maintaining cross guideline level. On a 1½″ grade, the pattern should now be a total of ⅝″ from the length guideline. Trace the wrist intersection and blend the overarm seam from the wrist to the shoulder notch, using the elbow notch as a guide.

5 WRIST GRADE

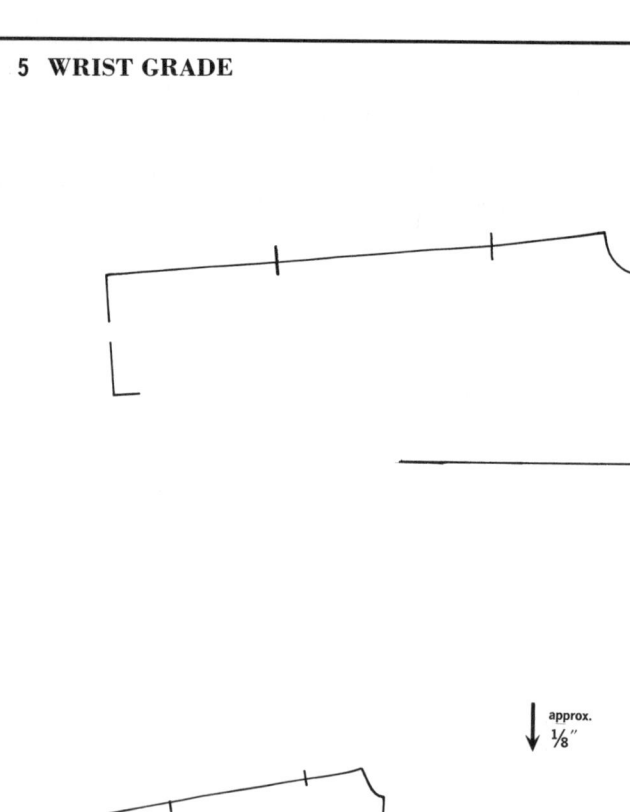

Shift the pattern down so that the cross control line of the pattern is 1/8" below the cross guideline on the paper. The total shift of the pattern from its previous position will be approximately 1/8". Trace the wrist-underarm seam intersection. Blend the wrist. For a straight sleeve, shift the pattern down 1/4".

7 UNDERARM GRADE

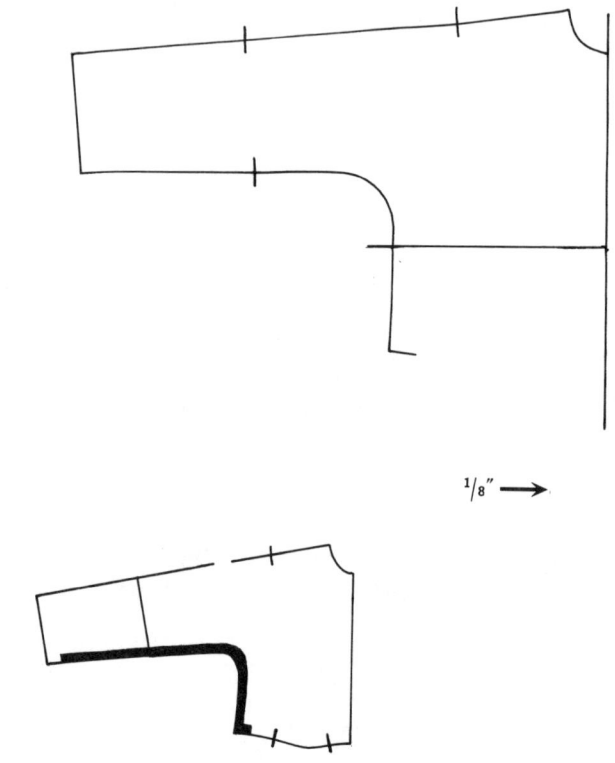

Shift the pattern in 1/8" from its previous position. On a 1½" grade, the pattern should now be a total of 3/8" from the length guideline. Trace the underarm-side seam intersection to and around the waistline intersection. Blend the underarm.

6 ELBOW GRADE

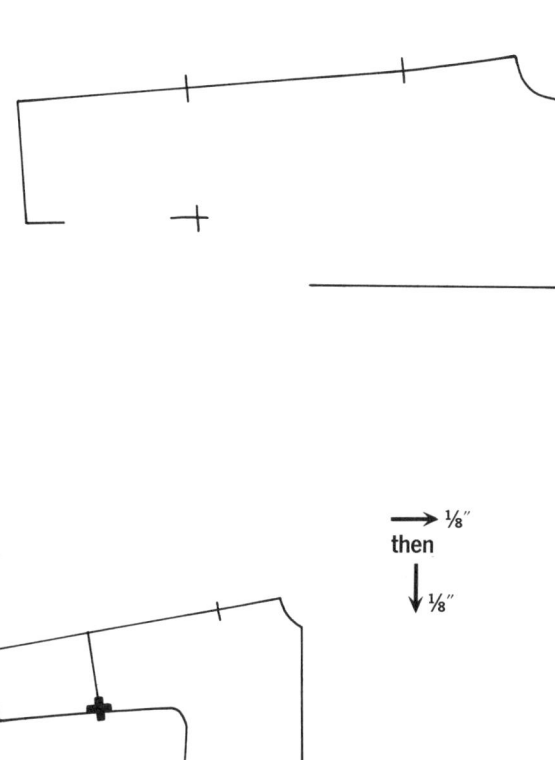

Shift the pattern in ⅛″ from its previous position. On a 1½″ grade, the pattern should now be a total of ½″ from the length guideline. Trace nothing.

Then shift the pattern down ⅛″ from its previous position. The cross control line of the pattern should now be a total of ¼″ below the cross guideline on the paper. Crossmark the elbow intersection. On a straight sleeve, omit the downward shift.

8 WAISTLINE GRADE

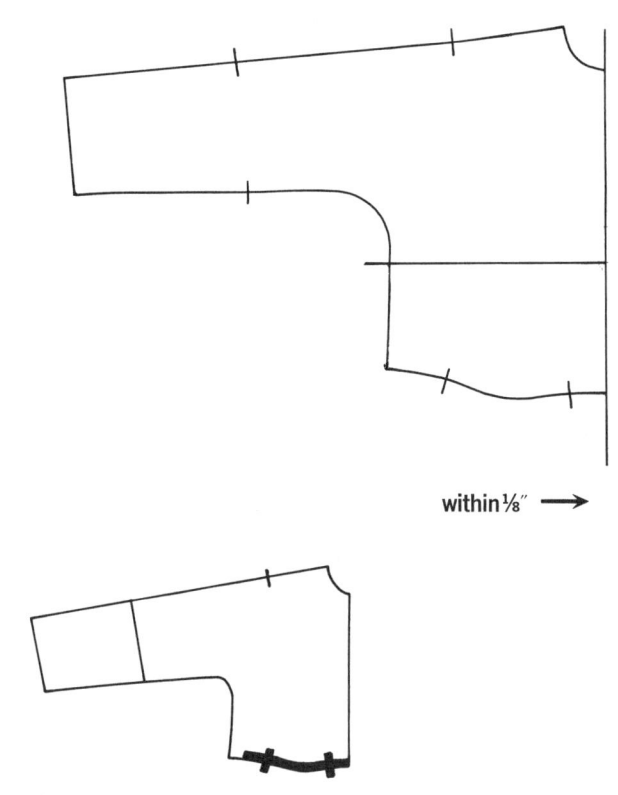

Shift the pattern in until it is within ⅛″ of the length guideline. Trace the waistline and the shirring notches. (For a pattern with a waistline dart follow steps 8 and 9 of the front bodice grade on page 27). Blend the waistline across to the length guideline, thus establishing the new Center Front-waistline intersection.

NOTE: The back kimono sleeve pattern is graded in a similar manner. For a fitted sleeve, the elbow dart is graded after the first shift of step 6 by marking the apex of the dart. After the second shift of step 6, mark the dart notches. The elbow dart is graded a standard ⅛″ per size on the kimono and gusset sleeves.

HOW THE BODY GROWS

KIMONO RAGLAN SLEEVE

1½″ Grade

1. The bodice is increased the same amount as a basic bodice, with the raglan seam receiving the combined armhole and shoulder increases of the basic bodice.

2. The split diagram of the kimono raglan sleeve pattern is shown here aligned with the raglan bodice pattern. The neck seam of the raglan sleeve is not graded in order to maintain the proportion created by the designer.

3. Maintaining shoulder level is not possible on a kimono raglan sleeve which has no shoulder seam. For this reason, the neck increase of ⅛″ per size which usually tapers to nothing at the elbow is here continued as a full ⅛″, and the entire elbow grade becomes a standard ½″ per size on the kimono raglan sleeve, rather than a variable grade as on the set-in sleeve.

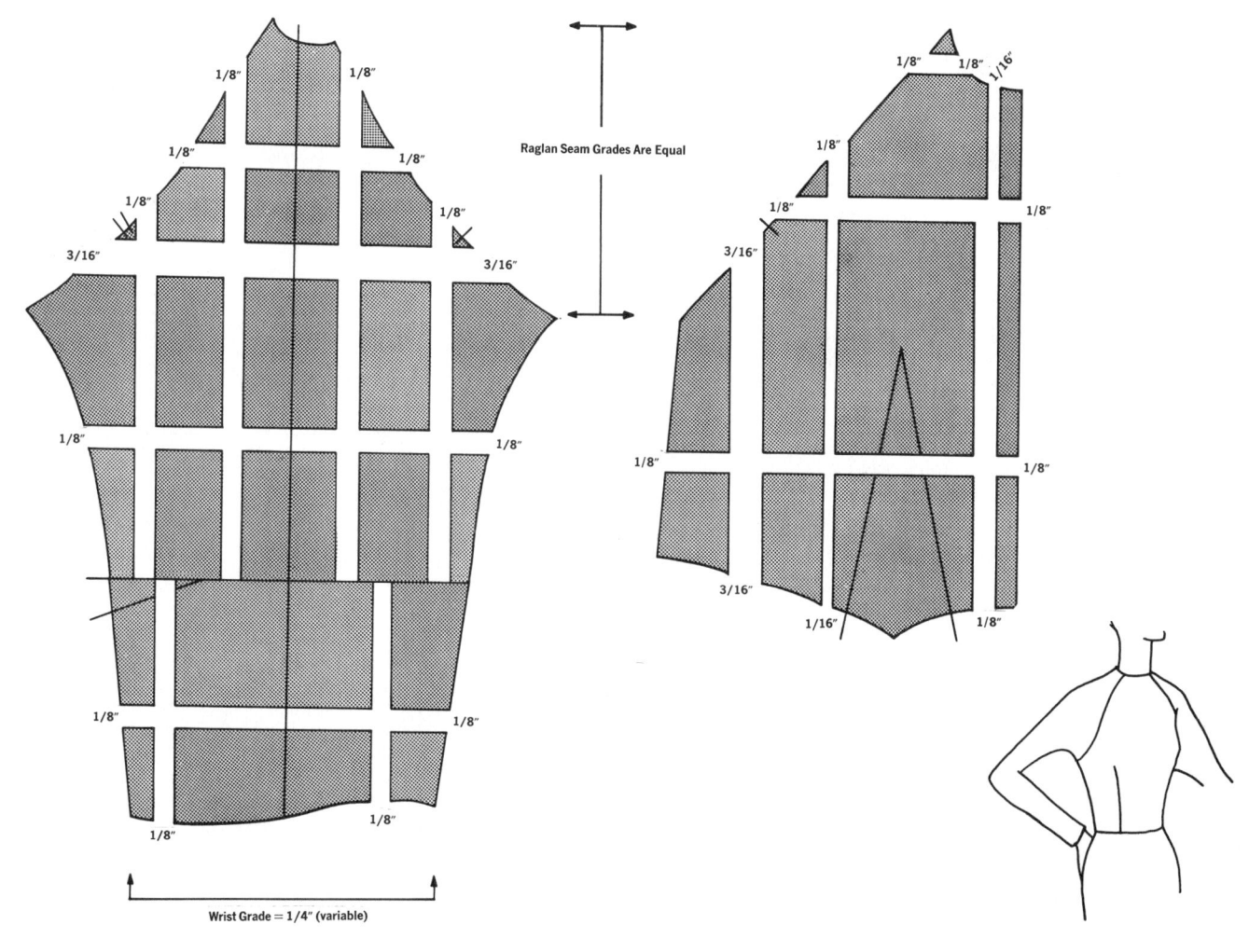

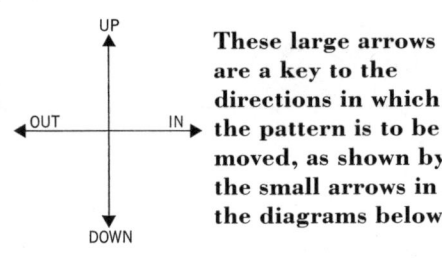

These large arrows are a key to the directions in which the pattern is to be moved, as shown by the small arrows in the diagrams below.

DIRECTIONS FOR A KIMONO RAGLAN SLEEVE
1½ INCH GRADE—BODICE SECTION

Pattern of Sleeve

Pattern of Front Bodice

1 CENTER FRONT

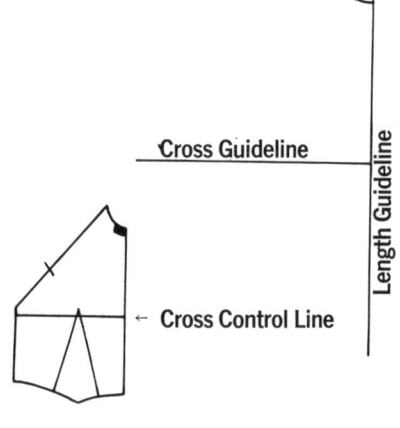

Cross Guideline
Length Guideline
← Cross Control Line

3 NECK GRADE

Shift the bodice out $1/16''$ from the length guideline, maintaining established raglan seam level. Trace the raglan seam-neckline intersection. Blend the neckline as shown on page 20.

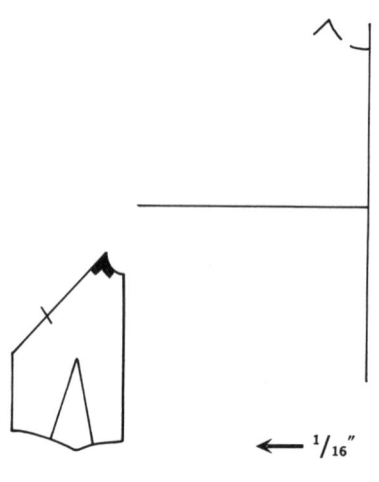

← $1/16''$

4 CROSS SHOULDER GRADE*

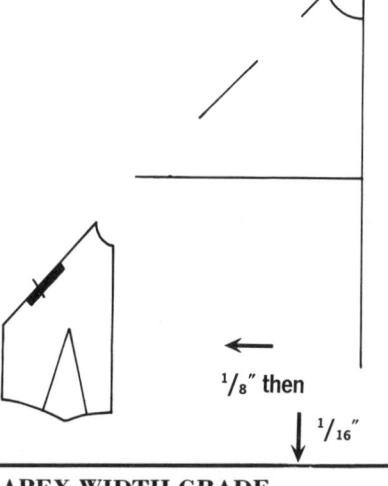

← $1/8''$ then
↓ $1/16''$

6 SIDE SEAM GRADE

Shift the bodice pattern down $1/8''$ from its previous position. The cross control line of the pattern should now be a total of $1/4''$ below the cross guideline on the paper. Trace the side seam to and around the waistline intersection.

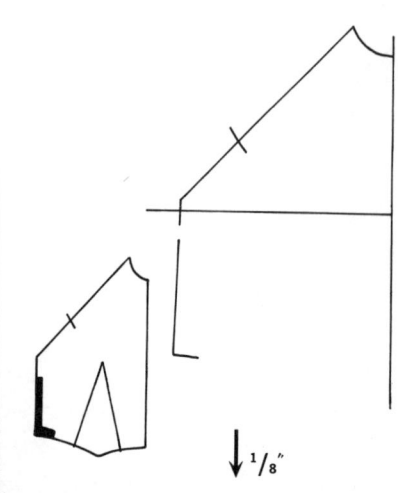

↓ $1/8''$

7 APEX WIDTH GRADE

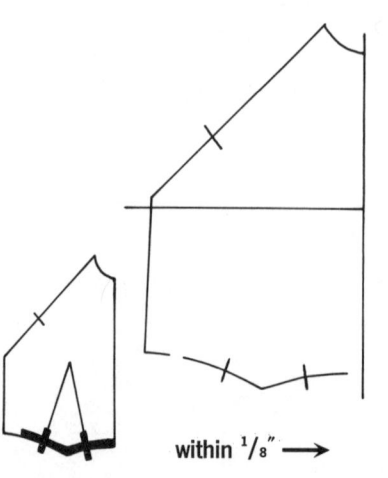

within $1/8''$ →

1. In all steps except the first shift of step 4* and the second shift of step 5*, the measurements used are standard from one size to the next no matter if it be a 1", 1½", or 2" grade. Refer to the bodice grading chart for the measurements for the first shift of step 4 and the second shift of step 5 *when doing other* than the 1½" grade shown below, and follow the directions given for the front bodice.

2. The smaller diagram in the lower corner of each box depicts the pattern being graded, with the section to be traced in that step marked in heavy lines.

Draw a cross control line on the pattern by squaring a line from Center Front to the side seam. On paper, draw a length guideline, and square a cross guideline from it. Place Center Front of the pattern on the length guideline, with the cross control line of the pattern on the cross guideline on the paper. Trace the neckline-Center Front intersection. (The length guideline will be the Center Front of the graded pattern).

2 SHOULDER LEVEL GRADE

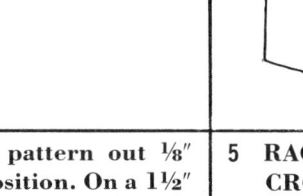

Shift the bodice pattern up on the length guideline, until the cross control line of the pattern is ⅛" above the cross guideline on the paper. Trace nothing.

Shift the bodice pattern out ⅛" from its previous position. On a 1½" grade, the pattern should now be a total of 3/16" from the length guideline. Trace nothing.

Then shift the pattern down 1/16" from its previous position. The cross control line of the pattern should now be a total of 1/16" above the cross guideline on the paper. Trace the middle of the raglan seam.

5 RAGLAN SEAM AND CROSS BUST GRADE*

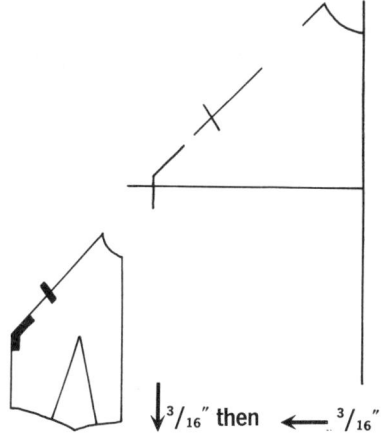

Shift the bodice pattern down so that the cross control line of the pattern is ⅛" below the cross guideline on the paper. The total downward shift of the pattern from its previous position will be 3/16". Trace nothing.

Then shift the pattern out 3/16" from its previous position. On a 1½" grade, the pattern should now be a total of ⅜" from the length guideline. Trace the lower raglan seam to and around the side seam intersection. Blend the raglan seam and mark the front notch.

Shift the bodice pattern in until it is within ⅛" of the length guideline. Trace the waistline, and mark the dart notches. Blend the waistline across to the length guideline, thus establishing the new Center Front-waistline intersection.

8 DART LENGTH GRADE

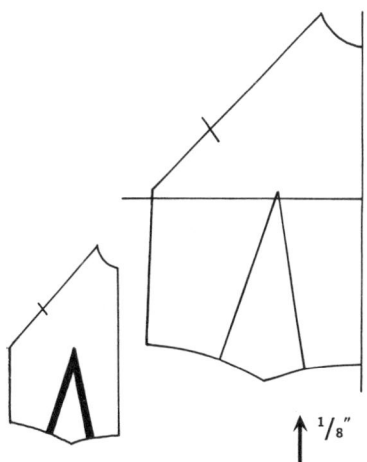

With the bodice pattern still within ⅛" of the length guideline, shift the pattern up ⅛" from its previous position and mark the dart apex. Lift the pattern off the paper and draw in the dartlines.

The back kimono raglan bodice pattern is graded in a similar manner.

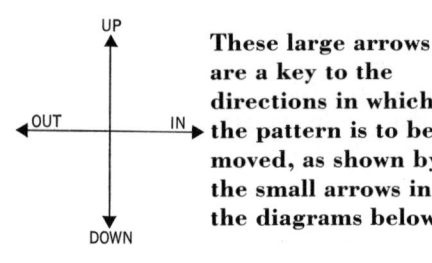

These large arrows are a key to the directions in which the pattern is to be moved, as shown by the small arrows in the diagrams below.

DIRECTIONS FOR A KIMONO RAGLAN SLEEVE 1½″ GRADE—SLEEVE SECTION

1 FRONT BICEP GRADE I

Draw a cross control line on the pattern by ruling a line across the bicep. Draw an elbow level line on the pattern. Draw a length control line on the pattern by ruling a line through the center of the sleeve. On paper, draw a length guideline and square a cross guideline from it. Place the pattern with its control lines on top of the corresponding guidelines on the paper. Shift the sleeve pattern in ¼″ from the length guideline. Trace the front bicep corner.

2 FRONT BICEP GRADE II

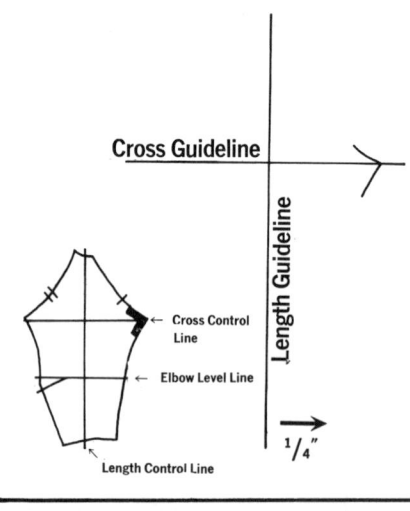

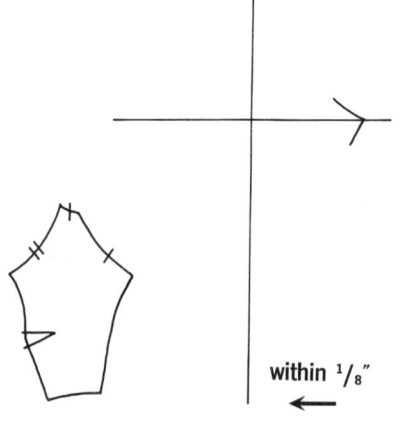

4 CROSS SHOULDER GRADE*

Shift the sleeve pattern up ⅛″ from its previous position. On a 1½″ grade, the cross control line on the pattern should now be a total of ⁵⁄₁₆″ above the cross guideline on the paper. Trace nothing.

5 FRONT RAGLAN GRADE

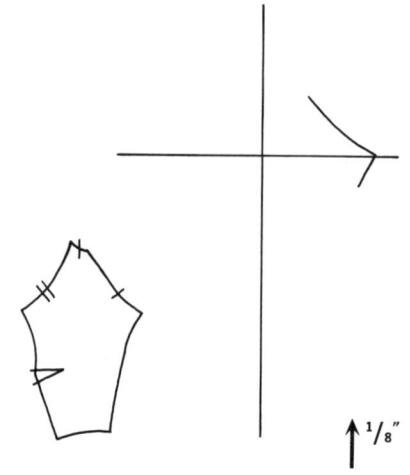

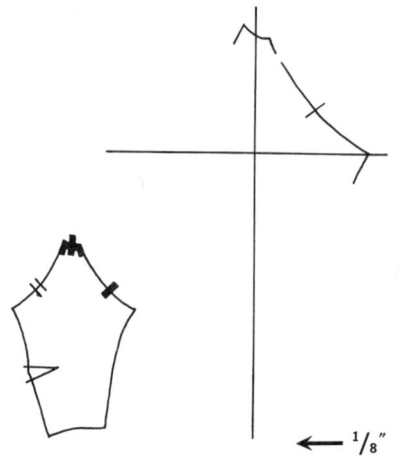

7 BACK SLEEVE CROSS BUST GRADE*

Shift the sleeve pattern down ³⁄₁₆″. The cross control line of the pattern will now lie on top of the cross guideline on the paper. Trace nothing.

8 BACK BICEP GRADE

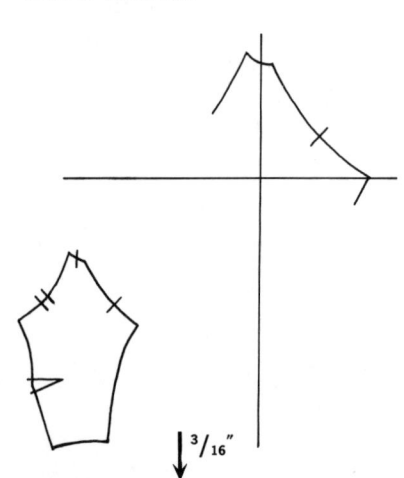

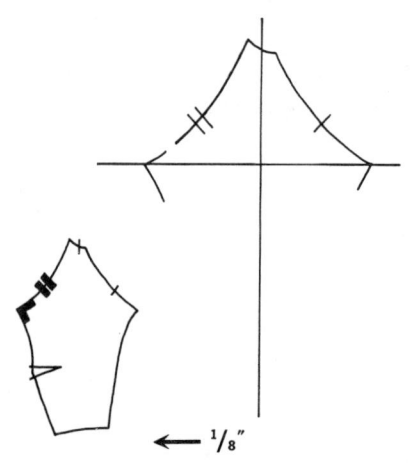

1. Steps 3* and 7* must always equal the second shift in step 5 of the front raglan bodice.
2. Steps 4* and second shift in step 6* must always equal step 4 in the front raglan bodice.
3. The smaller diagram in the lower corner of each box depicts the pattern being graded, with the section to be traced in that step marked in heavy lines.

Shift the sleeve pattern out until the length control line of the pattern is within ⅛″ of the length guideline on the paper. Trace nothing.

3 FRONT SLEEVE CROSS BUST GRADE*

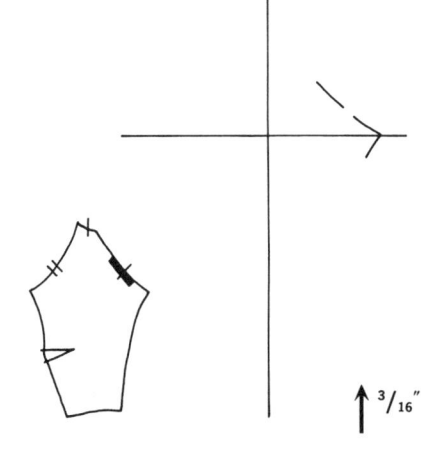

Shift the sleeve pattern up ³⁄₁₆″ from its previous position. Trace the middle of the front raglan seam.

Shift the sleeve pattern out ⅛″ so that the length control line of the pattern lies on top of the length guideline on the paper. Trace the remainder of the front raglan seam to and around the neck seam, to the back raglan seam intersection. Mark the shoulder notch. Blend the front raglan seam and mark the front sleeve notch.

6 BACK RAGLAN AND CROSS SHOULDER GRADE*

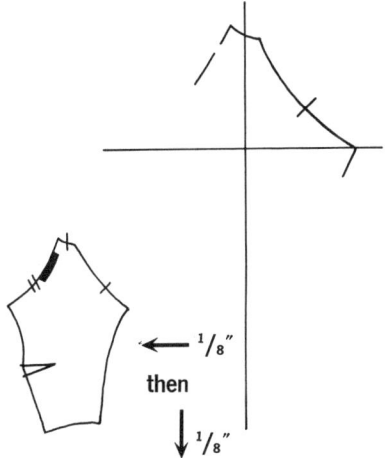

Shift the sleeve pattern out ⅛″ from the length guideline. Trace nothing.

Then shift the sleeve pattern down ⅛″. The cross control line of the pattern should be ³⁄₁₆″ above the cross guideline on the paper. Trace the middle of the back raglan seam.

Shift the sleeve pattern out ⅛″ from its previous position. The cross control line of the pattern should now be a total of ¼″ from the cross guideline on the paper. Trace the back bicep corner. Blend the back raglan seam and trace the back notches.

9 REMAINDER OF SLEEVE

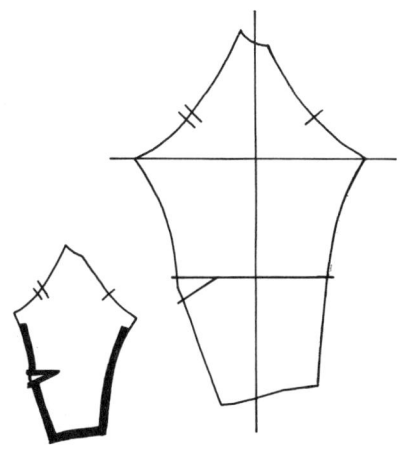

For the remainder of the sleeve follow the directions for the set-in sleeve on page 5 starting at step 6. For a straight sleeve the wrist is graded the same amount as the elbow.

HOW THE BODY GROWS

SQUARE ARMHOLE | 1½" Grade

1. In order to preserve the designer's proportions, the square-designed style lines of the bodice and of the sleeve remain the same length for all sizes.

2. A fitted sleeve wrist is graded one-half of the bicep grade, while on a straight sleeve, as in the diagram below, the wrist is graded the same as the bicep.

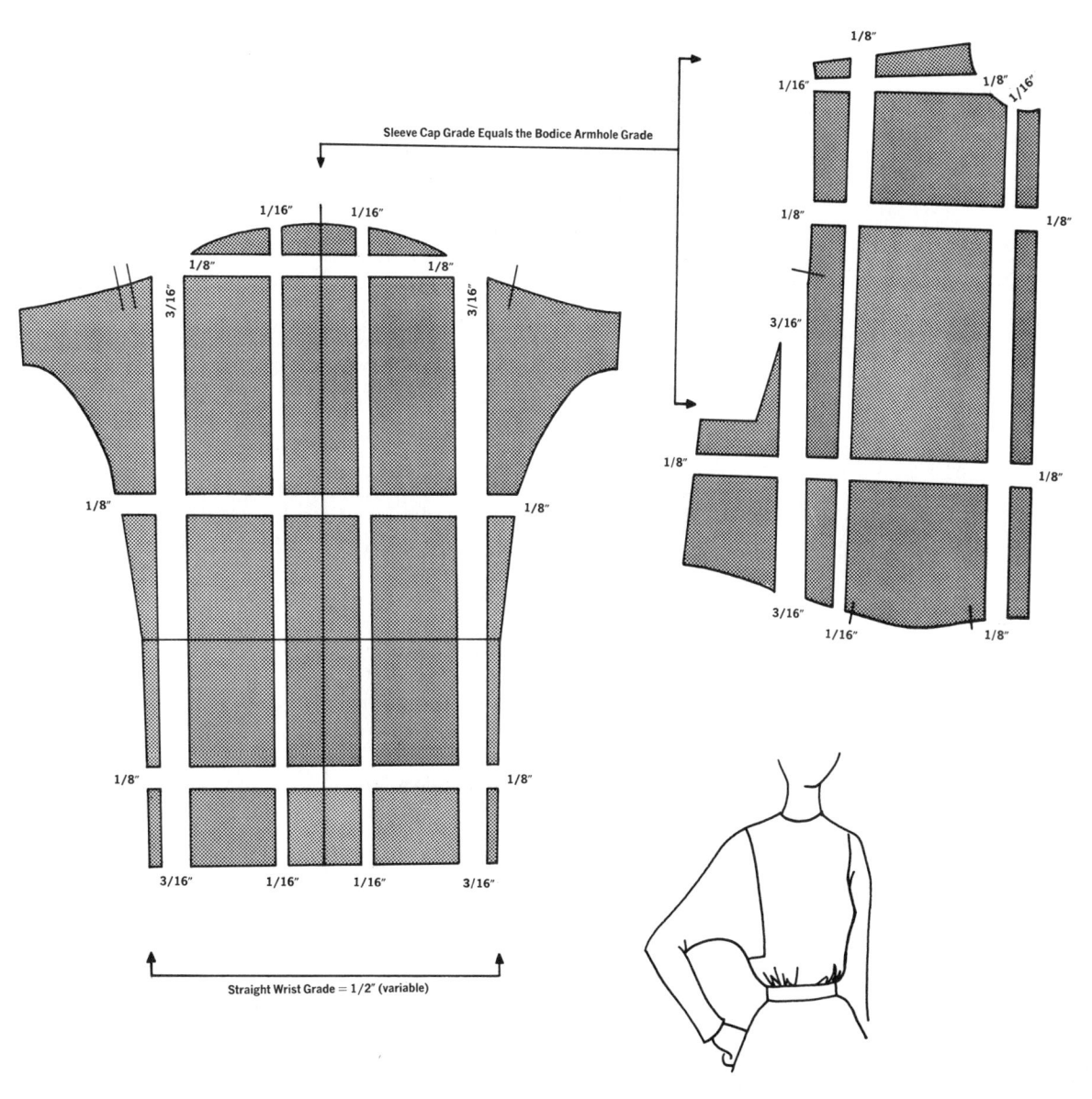

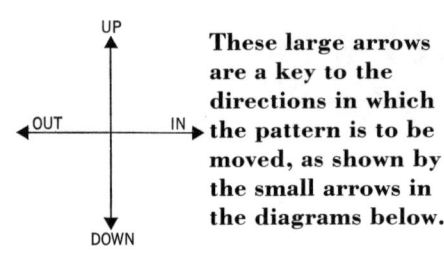

These large arrows are a key to the directions in which the pattern is to be moved, as shown by the small arrows in the diagrams below.

DIRECTIONS FOR A SQUARE ARMHOLE
1½″ GRADE—BODICE SECTION

Pattern of Front Bodice

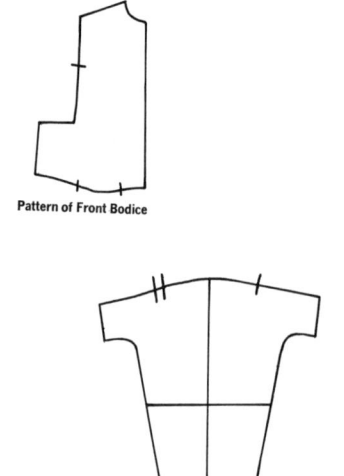

Pattern of Sleeve

1 CENTER FRONT

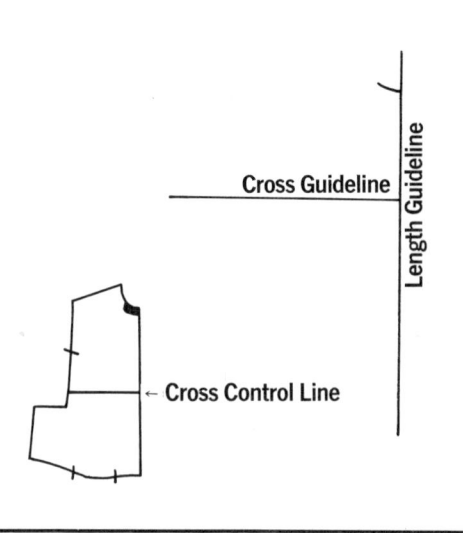

3 NECK GRADE

← 1/16″

Shift the pattern out 1/16″ from the length guideline. Before tracing, adjust the pattern so that the shoulder seam is once again directly on the shoulder level traced in step 2. This adjustment is known as "maintaining shoulder level." Now trace the shoulder—neckline intersection. Blend the neckline as shown on page 20.

4 CROSS SHOULDER GRADE*

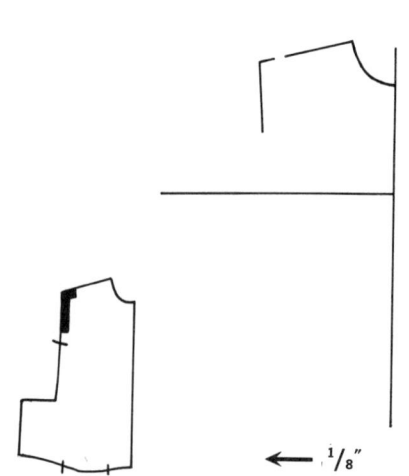
← 1/8″

6 CROSS BUST GRADE*

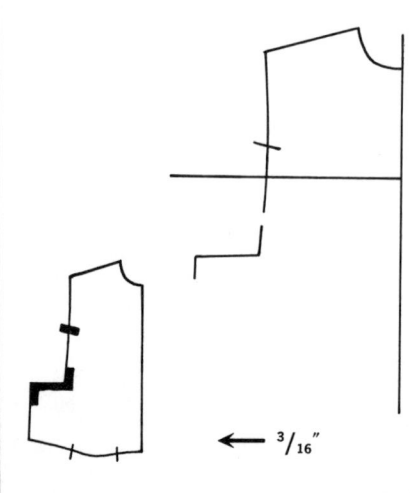
← 3/16″

Shift the pattern out 3/16″ from its previous position. The pattern should now be a total of 3/8″ from the length guideline. Trace the lower armhole, style line, and side seam intersection. Blend the armhole as shown on page 21, and mark the armhole notch.

7 SIDE SEAM GRADE

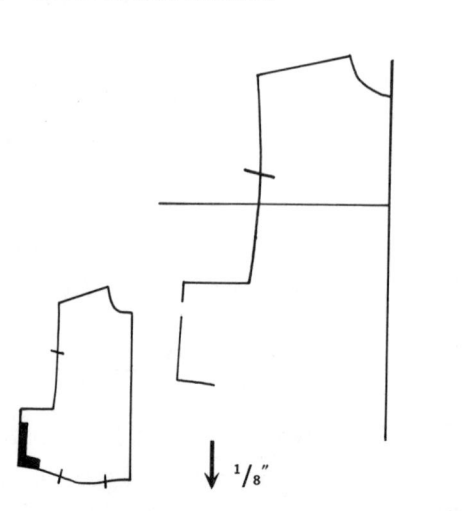
↓ 1/8″

1. In all steps except 4* and 6*, the measurements used are standard from one size to the next, no matter if it be a 1", 1½" or 2" grade. Refer to the bodice grading chart for measurements for steps 4 and 6 *when doing other* than the 1½" grade shown below. For step 4, subtract column 3 from column 4 before applying the measurement. For step 6, divide the measurement in column 6 in half before using.

2. The smaller diagram in the lower corner of each box depicts the pattern being graded, with the section to be traced in that step marked in heavy lines.

Draw a cross control line on the pattern by squaring a line from the middle of Center Front. On paper, draw a length guideline, and square a cross guideline from it. Place Center Front of the pattern on the length guideline, with the cross control line of the pattern on the cross guideline. Trace the neckline—Center Front intersection. (The length guideline will be the Center Front of the graded pattern.)

2 SHOULDER LEVEL GRADE

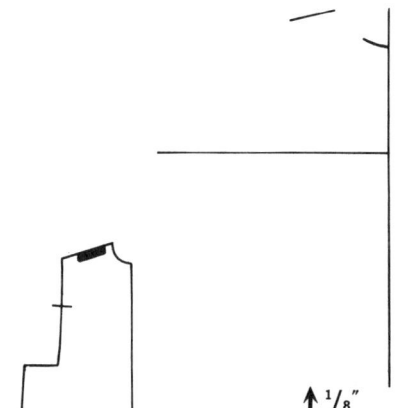

↑ ⅛"

Shift the pattern up on the length guideline, until the cross control line of the pattern is ⅛" above the cross guideline on the paper. Trace the middle of the shoulder seam in order to establish the shoulder level.

Shift the pattern out ⅛" from its previous position, maintaining shoulder level. The pattern should now be a total of 3/16" from the length guideline. Trace the remainder of the shoulder, and one third of the armhole. Blend the shoulder seam as shown on page 19. (See note, Step 4, Front Bodice).

5 ARMHOLE GRADE

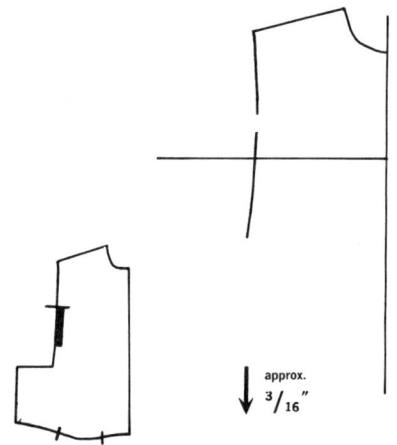

↓ approx. 3/16"

Shift the pattern down so that the cross control line of the pattern is ⅛" below the cross guideline on the paper. The total downward shift of the pattern from its previous position will be approximately 3/16". Trace the middle of the armhole.

Shift the pattern down ⅛" from its previous position. The cross control line of the pattern should now be a total of ¼" below the cross guideline of the paper. Trace the side seam to and around the waistline intersection.

8 APEX WIDTH GRADE

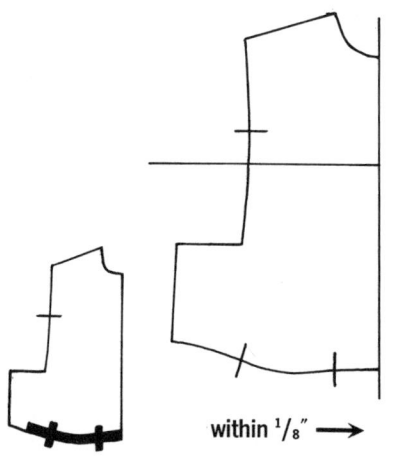

within ⅛" →

Shift the pattern in until it is within ⅛" of the length guideline. Trace the waistline, and mark the shirring notches. Blend the waistline across to the length guideline, thus establishing the new Center Front—waistline intersection. The square armhole back bodice is graded in a similar manner.

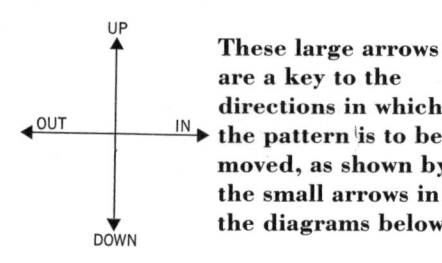

These large arrows are a key to the directions in which the pattern is to be moved, as shown by the small arrows in the diagrams below.

DIRECTIONS FOR A SQUARE ARMHOLE
1½" GRADE—STRAIGHT SLEEVE SECTION

1 BICEP GRADE*

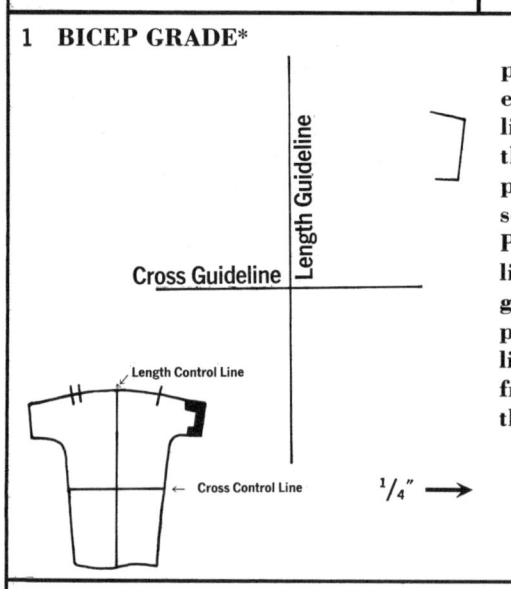

Draw a cross control line on the pattern by drawing a line across the elbow level. Draw a length control line on the pattern by ruling a line through the center of the sleeve. On paper, draw a length guideline and square a cross guideline from it. Place the pattern with its control lines on top of the corresponding guidelines on the paper. Shift the pattern in ¼" from the length guideline. Trace the front square armhole from the bicep corner to and around the cap corner.

2 FRONT CAP GRADE

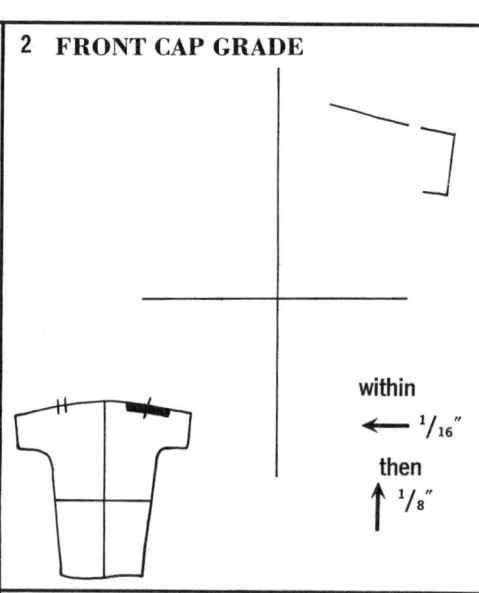

within ← 1/16"
then ↑ 1/8"

4 CAP WIDTH GRADE—Back

Shift the pattern out 1/16" from the length guideline. Trace the top and middle of the cap.

← 1/16"

5 BACK CAP GRADE*

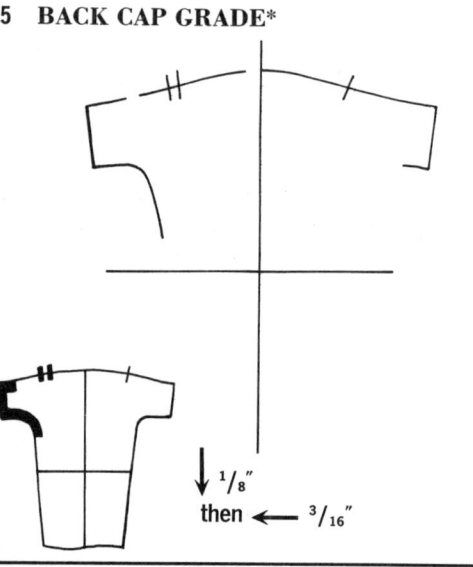

↓ 1/8"
then ← 3/16"

7 BACK WRIST GRADE

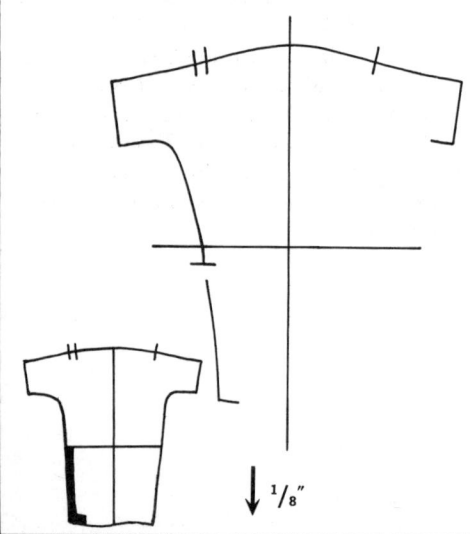

Shift the pattern down 1/8" from its previous position. The cross control line of the pattern should now be a total of ¼" below the cross guideline on the paper. Trace the back underarm seam from the elbow level to the wrist intersection.

↓ 1/8"

8 FRONT WRIST GRADE

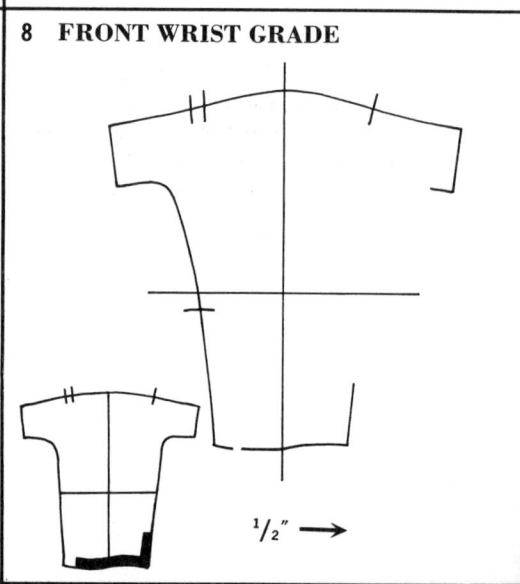

½" →

1. In all steps except 1* and the second shift of step 5*, the measurements used are standard from one size to the next, no matter if it be a 1″, 1½″, or 2″ grade. Refer to column 1 of the sleeve grading chart for the measurements for steps 1 and the second shift of step 5 *when doing other than the 1½″ grade shown below.* For step 1, divide the measurement in half before using. For the second shift of step 5, divide the measurement in half and subtract ¹⁄₁₆″ from that measurement before using.

2. The smaller diagram in the lower corner of each box depicts the pattern being graded, with the section to be traced in that step marked in heavy lines.

Shift the pattern out until the length control line of the pattern is within ¹⁄₁₆″ of the length guideline on the paper. Trace nothing.

Then shift the pattern up until the cross control line of the pattern is ⅛″ above the cross guideline on the paper. Trace the middle of the cap.

3 CAP WIDTH GRADE—front

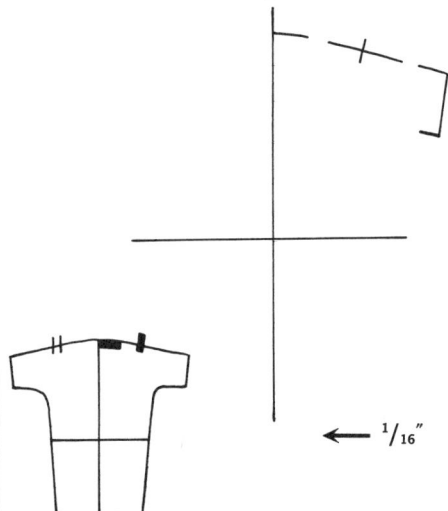

Shift the pattern out ¹⁄₁₆″ until the length control line of the pattern lies on top of the length guideline on the paper. Trace the top of the cap. Blend the sleeve cap and mark the front cap notch.

Shift the pattern down ⅛″ until the cross control line of the pattern lies on top of the cross guideline on the paper. Trace nothing.

Then shift the pattern out ³⁄₁₆″ from its previous position. On a 1½″ grade, the length control line of the pattern should now be a total of ¼″ out from the length guideline on the paper. Trace the back square armhole from the cap corner to and around the bicep corner and the underarm curve. Blend the sleeve cap and mark the back cap notches.

6 UNDERARM GRADE—back

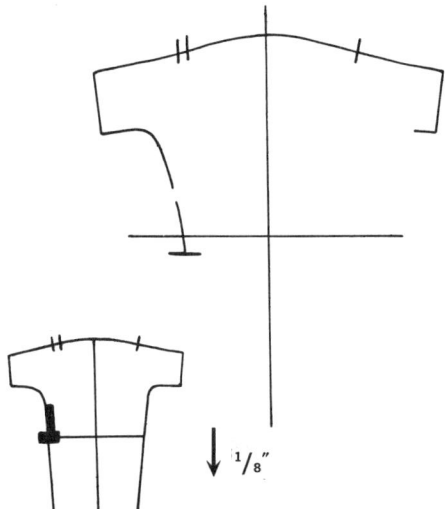

Shift the pattern down until the cross control line of the pattern is ⅛″ below the cross guideline on the paper. Trace the back underarm seam to the elbow level line, and mark an elbow level crossmark.

Shift the pattern in ½″ from its previous position. The length control line of the pattern should now be a total of ¼″ in from the length guideline on the paper. Trace the wrist to and around the front underarm seam intersection. Blend the wrist line.

NOTE: The inward shift of this step on a straight sleeve is always equal to the total bicep grade of the sleeve. For a 1″ grade, shift in ⅜″. For a 2″ grade, shift in ⅝″.

9 UNDERARM GRADE—front

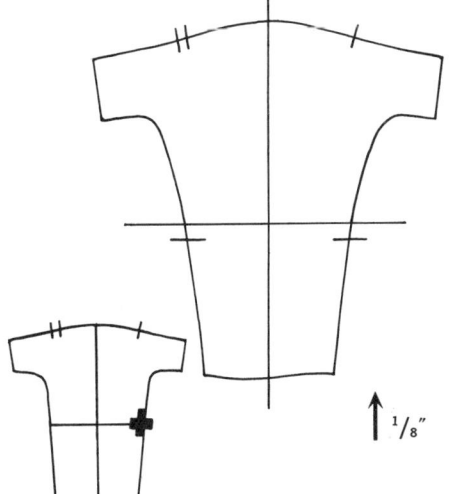

Shift the pattern up ⅛″ so that the cross control line of the pattern is ⅛″ below the cross guideline on the paper. Mark the elbow level crossmark. Blend a line from the front bicep corner to the front wrist-underarm intersection, using the elbow crossmark as a guide. The square armhole sleeve 1½″ grade is now complete.

HOW THE BODY GROWS

GUSSET SLEEVE

1½" Grade

1. The usual side seam grade of ⅛" per size is here graded *above* the actual side seam, as the grade is needed to achieve the proper width at the elbow.

2. The gusset seams and slashline are graded automatically within the bodice; the gusset seams must be measured by ruler in order to grade the diamond gusset itself.

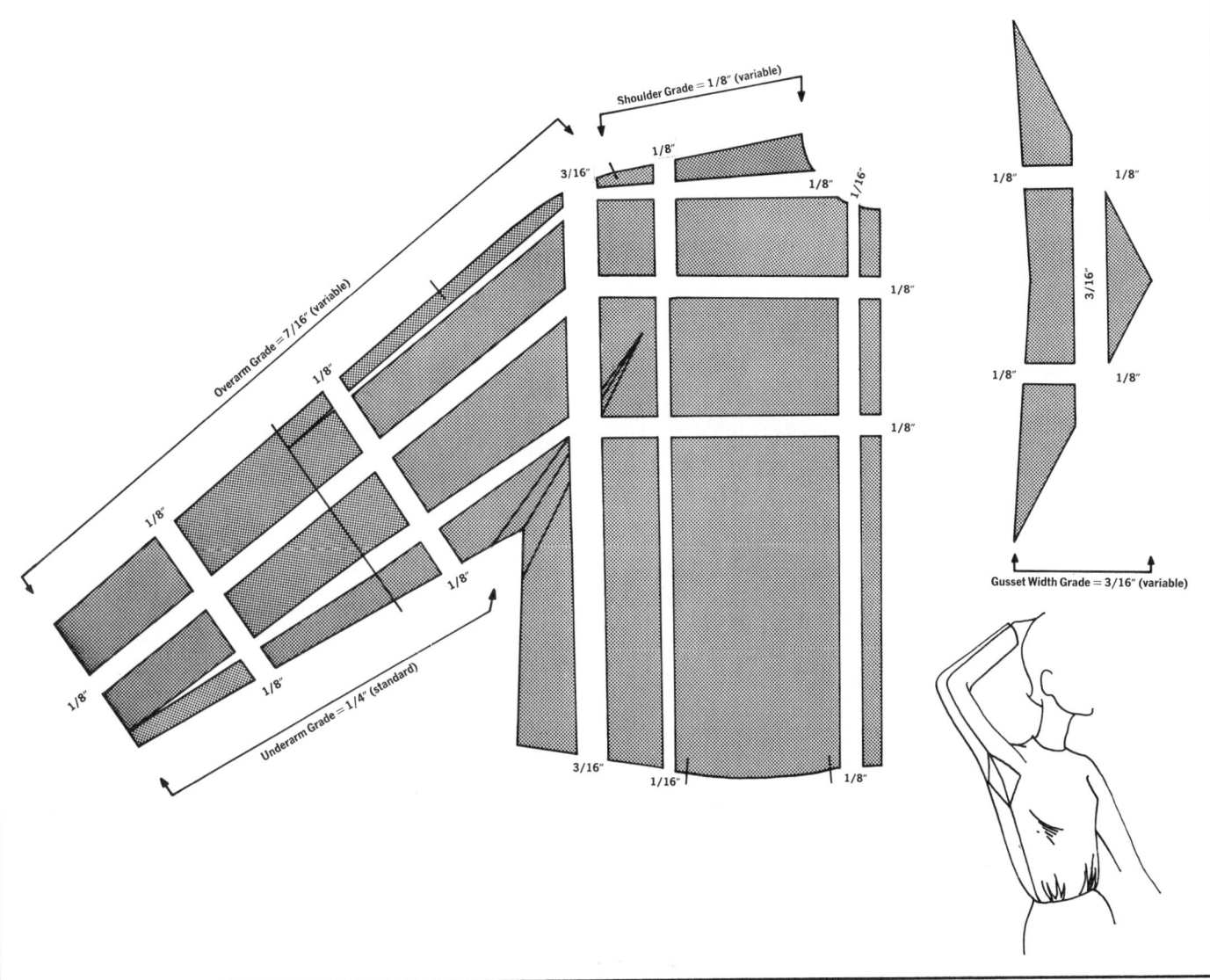

101

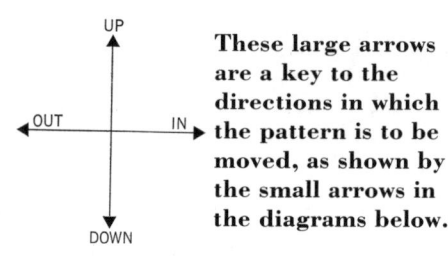

These large arrows are a key to the directions in which the pattern is to be moved, as shown by the small arrows in the diagrams below.

DIRECTIONS FOR A FRONT GUSSET SLEEVE 1½" GRADE

1 SHOULDER LEVEL GRADE

Draw a cross control line on the pattern by squaring a line from Center Front through the side seam and onto the sleeve. On paper, draw a length guideline, and square a cross guideline from it. Place Center Front of the pattern on the length guideline, with the cross control line of the pattern on the cross guideline on the paper. Trace the neckline—Center Front intersection. (The length guideline will be the Center Front of the graded pattern).

Shift the pattern up ⅛" on the length guideline. Trace the middle of the shoulder seam in order to establish the shoulder level.

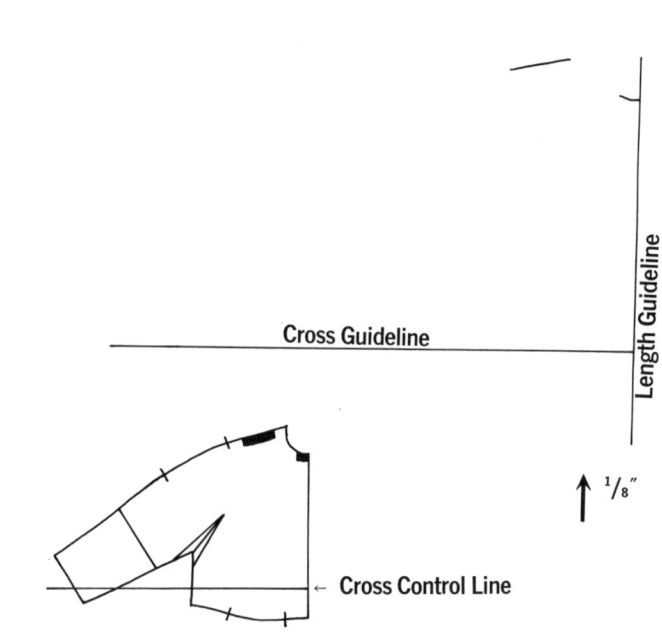

3 CROSS SHOULDER GRADE*

Shift the pattern out ⅛" from its previous position, maintaining shoulder level. On a 1½" grade, the pattern should now be a total of 3/16" from the length guideline. Trace the remainder of the shoulder to just past the shoulder notch. Mark the shoulder notch. (See note, step 4, Front Bodice).

Then shift the pattern down until the cross control line of the pattern lies on top of the cross guideline on the paper. The total downward shift of the pattern from its previous position will be approximately 1/16". Mark nothing.

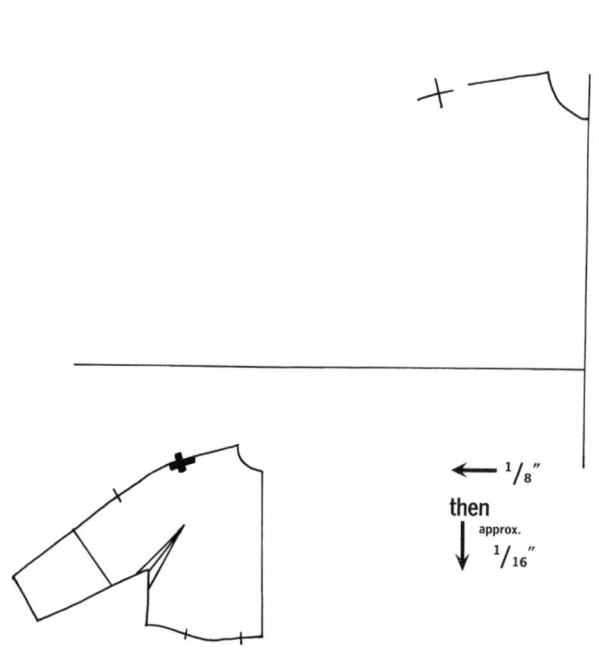

1. In all steps except the first shift of step 3,* and step 4,* the measurements used are standard from one size to the next, no matter if it be a 1″, 1½″, or 2″ grade. Refer to the grading chart on page 15 for the measurements for the first shift of step 3, and step 4 *when doing other* than the 1½″ grade shown below. For the first shift of step 3, subtract column 3 from column 4 before applying the measurement. For step 4, divide the measurement in column 6 in half before using.

2. The smaller diagram in the lower corner of each box depicts the pattern being graded, with the section to be traced in that step marked in heavy lines.

2 NECK GRADE

Shift the pattern out 1/16″ from the length guideline, maintaining shoulder level. Trace the shoulder—neckline intersection. Blend the neckline as shown on page 20.

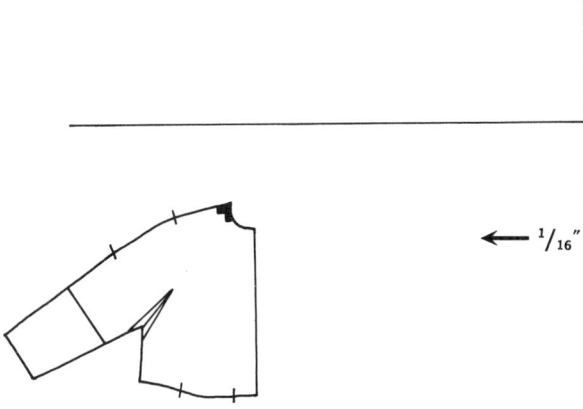

← 1/16″

4 CROSS BUST GRADE*

Shift the pattern out 3/16″ from its previous position maintaining cross guideline level. On a 1½″ grade, the pattern should now be a total of 3/8″ from the length guideline. Crossmark the bicep intersection.

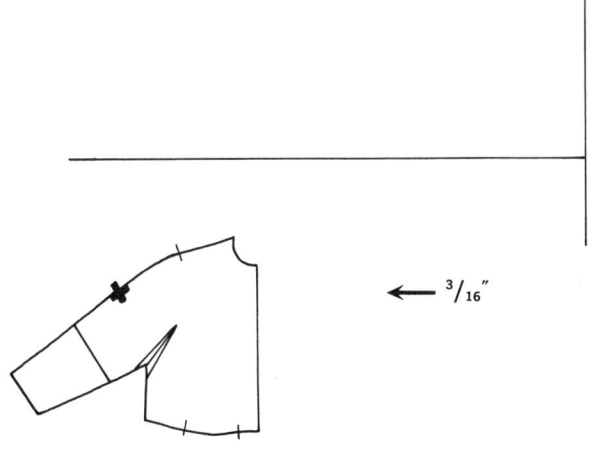

← 3/16″

103

5 OVERARM GRADE—I

DIRECTIONS FOR A FRONT GUSSET SLEEVE 1½" GRADE —CONTINUED

Shift the pattern out ⅛" from its previous position, maintaining cross guideline level. On a 1½" grade, the pattern should now be a total of ½" from the length guideline. Crossmark the elbow level intersection.

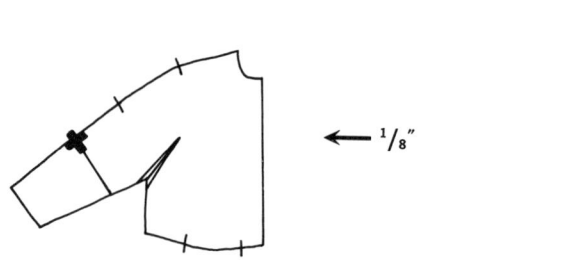

7 WRIST GRADE

Shift the pattern down so that the cross control line of the pattern is ⅛" below the cross guideline on the paper. The total shift of the pattern from its previous position will be ⅛". Trace the wrist—underarm seam intersection. Blend the wrist.

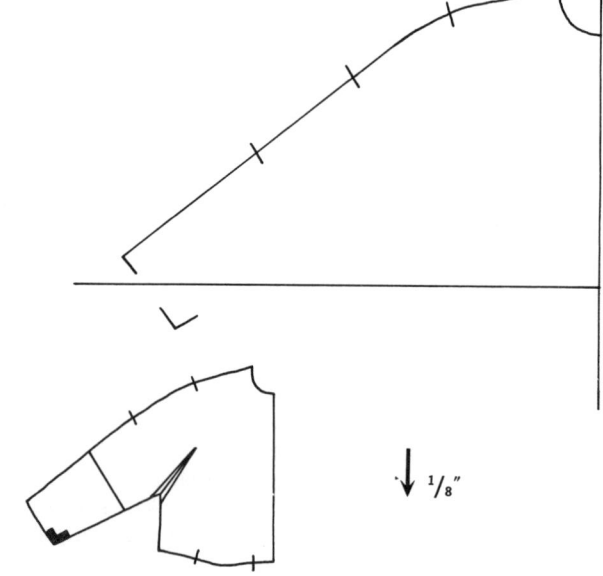

9 UNDERARM GRADE

Shift the pattern in ⅛" from its previous position. On a 1½" grade, the pattern should now be a total of ⅜" from the length guideline. Trace the underarm—side seam intersection to and around the waistline intersection. Blend the underarm seam and mark the gusset seam notches.

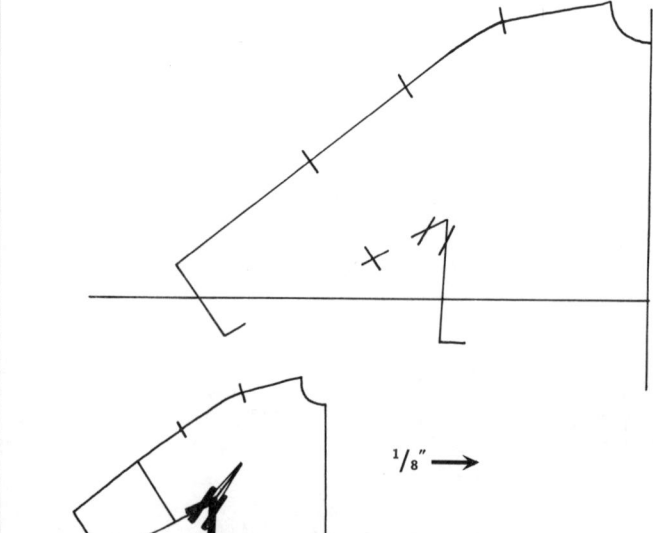

6 OVERARM GRADE—II

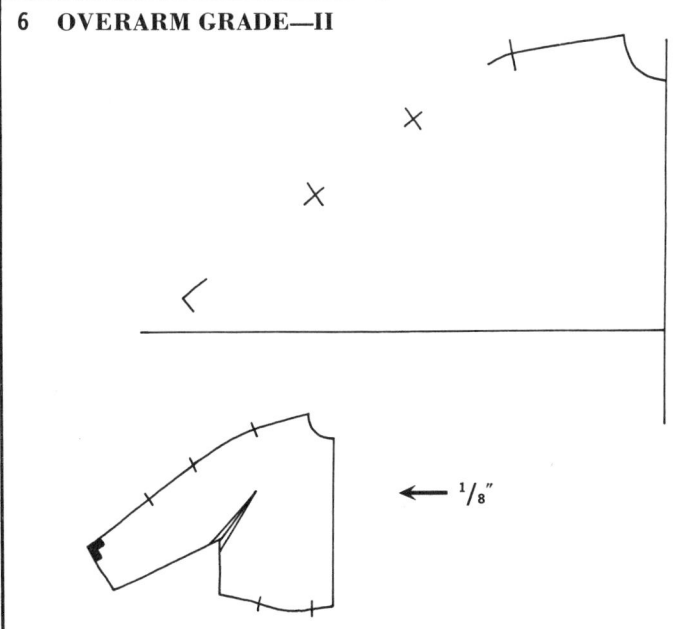

Shift the pattern out ⅛" from its previous position, maintaining cross guideline level. On a 1½" grade, the pattern should now be a total of ⅝" from the length guideline. Trace the overarm—wrist intersection, and blend the overarm seam.

8 ELBOW GRADE

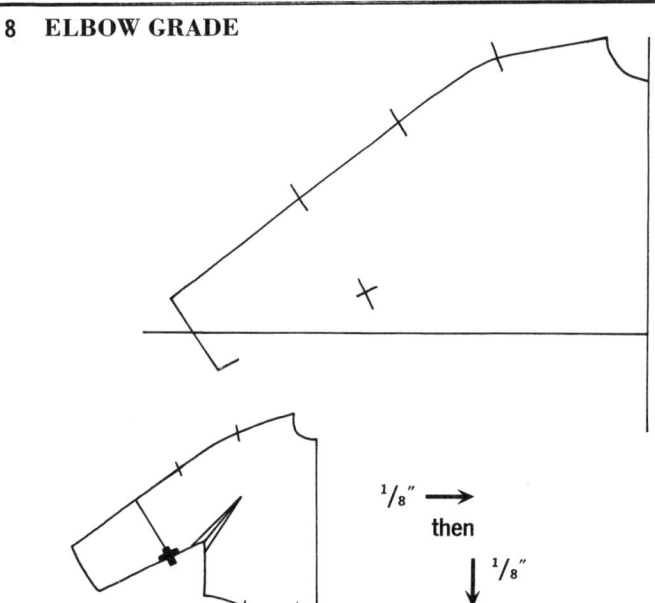

Shift the pattern in ⅛" from its previous position. On a 1½" grade, the pattern should now be a total of ½" from the length guideline. Trace nothing.

Then shift the pattern down ⅛" from its previous position. The cross control line of the pattern should now be a total of ¼" below the cross guideline on the paper. Crossmark the elbow intersection.

10 APEX WIDTH GRADE

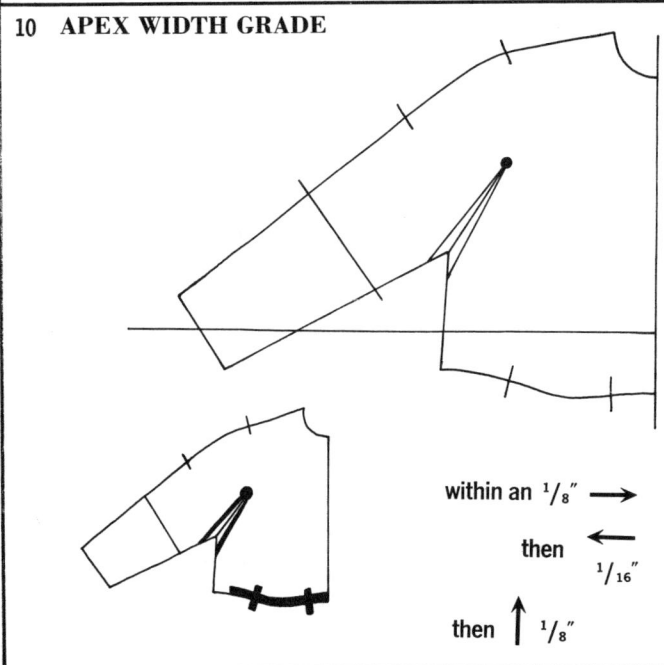

Shift the pattern in until it is within ⅛" of the length guideline. Trace the waistline and the shirring notches. For a pattern with a waistline dart follow steps 8 and 9 of the front bodice grade on page 27. Blend the waistline across to the length guideline, thus establishing the new Center Front—waistline intersection.

NOTE: After shifting in within an ⅛" and completing the necessary tracing, shift the pattern as follows:

For a 1" grade, shift up ⅛".
For a 1½" grade, shift out 1/16" and up ⅛".
For a 2" grade, shift out ⅛" and up ⅛".

Mark the gusset slash point. Lift the pattern off the paper and draw in the gusset seam lines to the slash point, and draw in the elbow line. Measure the length of one of the gusset seam lines and follow the instructions for the gusset on the following pages. A back gusset sleeve is graded in a similar manner. For the elbow dart, see the note of step 8 of the kimono sleeve lesson.

11 GUSSET PREPARATION

Draw a cross control line across the center of the gusset pattern. On paper, draw a length guideline, and square a cross guideline from it. Place the cross control line of the pattern on the cross guideline, with the ends of the gusset pattern touching the length guideline. Trace a short section of the inside center seam of the gusset, above and below the cross guideline.

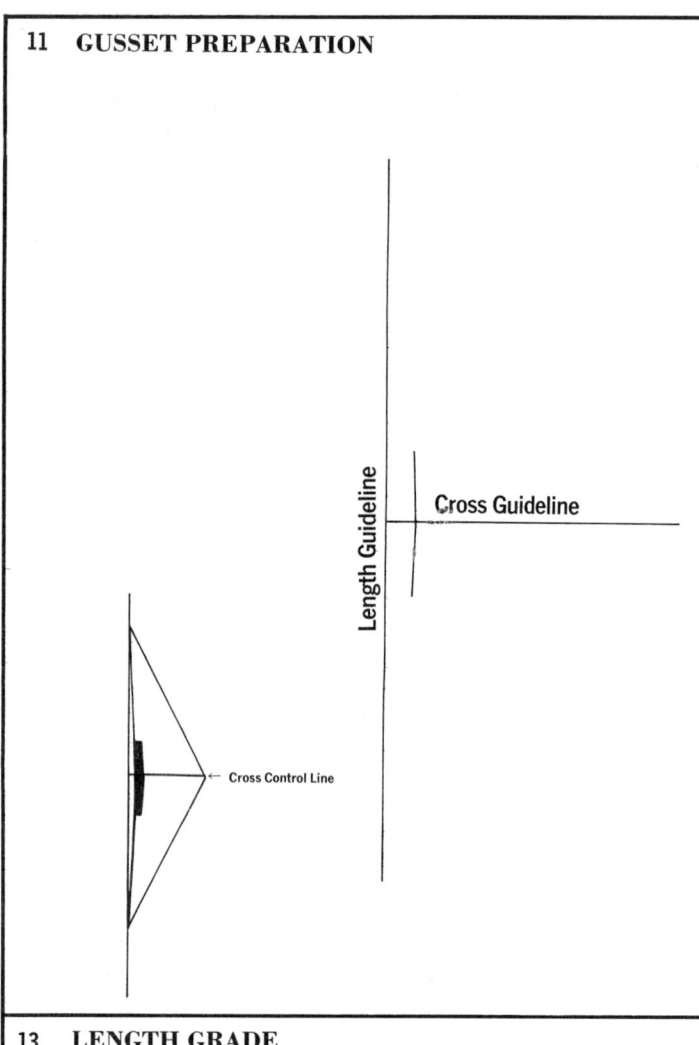

13 LENGTH GRADE

Lift off the pattern. Place a ruler on the outside point of the gusset at the cross guideline, and touch the ruler up to the length guideline at a point on the ruler equal to the length of the gusset seam line as measured on step 10, page 105. Draw a straight line. Repeat this step for the lower part of the gusset.

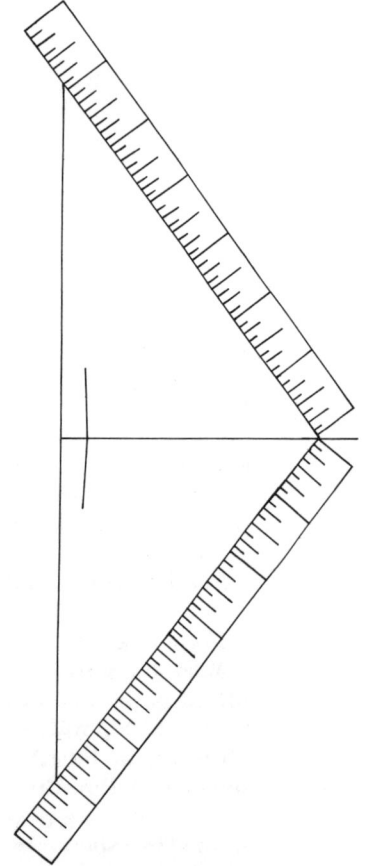

12 WIDTH GRADE

Shift the pattern in ³⁄₁₆" from the previously traced inside center seam. This is the same measurement as the cross bust grade of the gusset bodice on page 103, step 4. Therefore it would be shifted in ⅛" for a 1" grade, and ¼" for a 2" grade. Trace the outside point of the gusset for a short distance above and below the cross guideline.

14 GUSSET COMPLETION

Blend the remainder of the inside center seam of the gusset from the top and bottom points to the section of the inside center seams traced in step 11.

For a one piece diamond-shaped gusset follow the preceding steps, omitting in step 11 the tracing of the inside center seam.

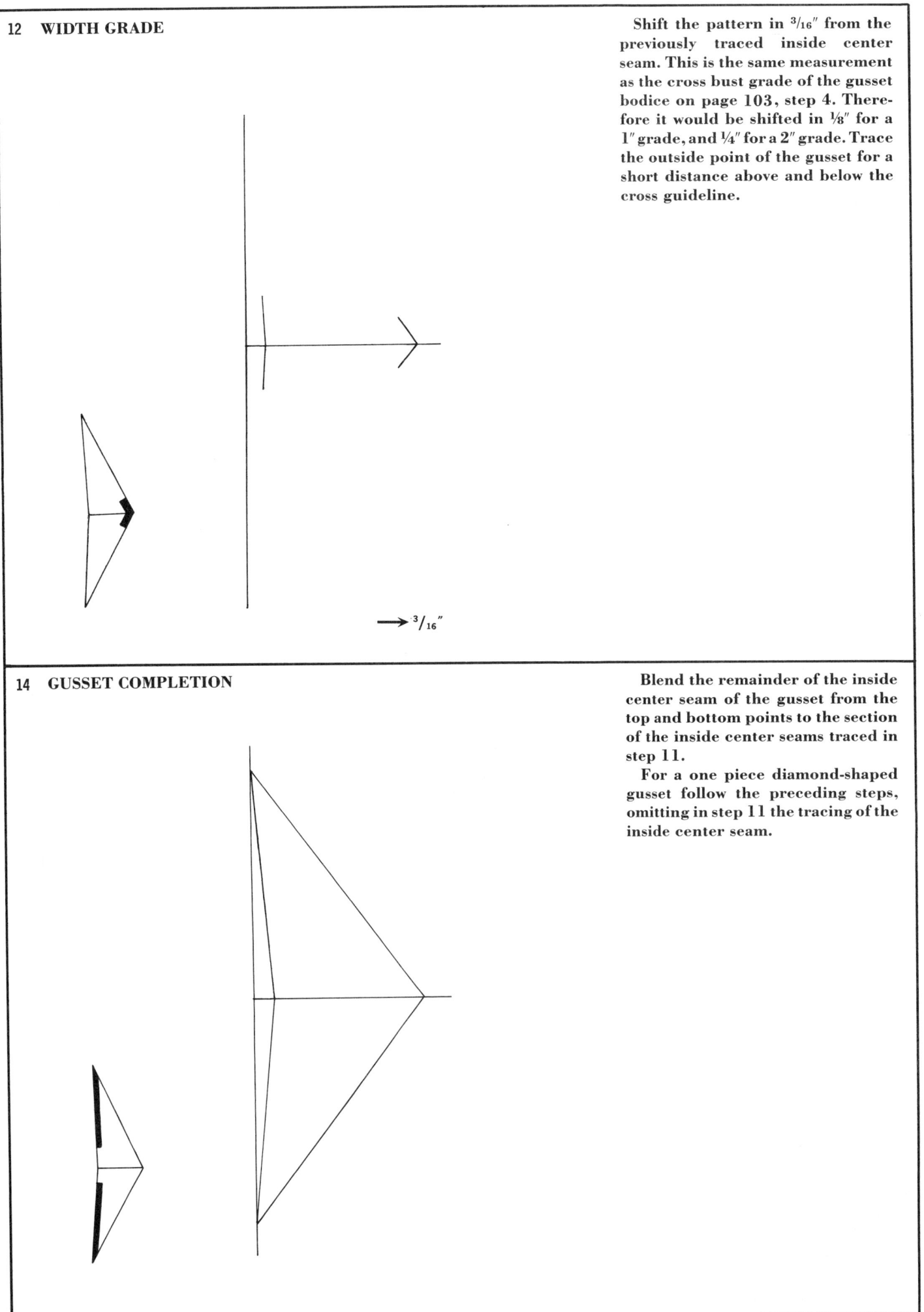

107

HOW THE BODY GROWS

SET-IN RAGLAN SLEEVE

1½" Grade

1. The bodice is increased the same amount as a basic bodice, with the raglan seam receiving the armhole and shoulder increases of the basic bodice.

2. The split diagram of the set-in raglan sleeve pattern is shown here aligned with the raglan bodice pattern. The raglan seam of the sleeve also receives the armhole and shoulder increases of the basic bodice. The overarm seam of the sleeve receives the shoulder and cap increases of the basic bodice and sleeve.

3. A short sleeve is not graded in length.

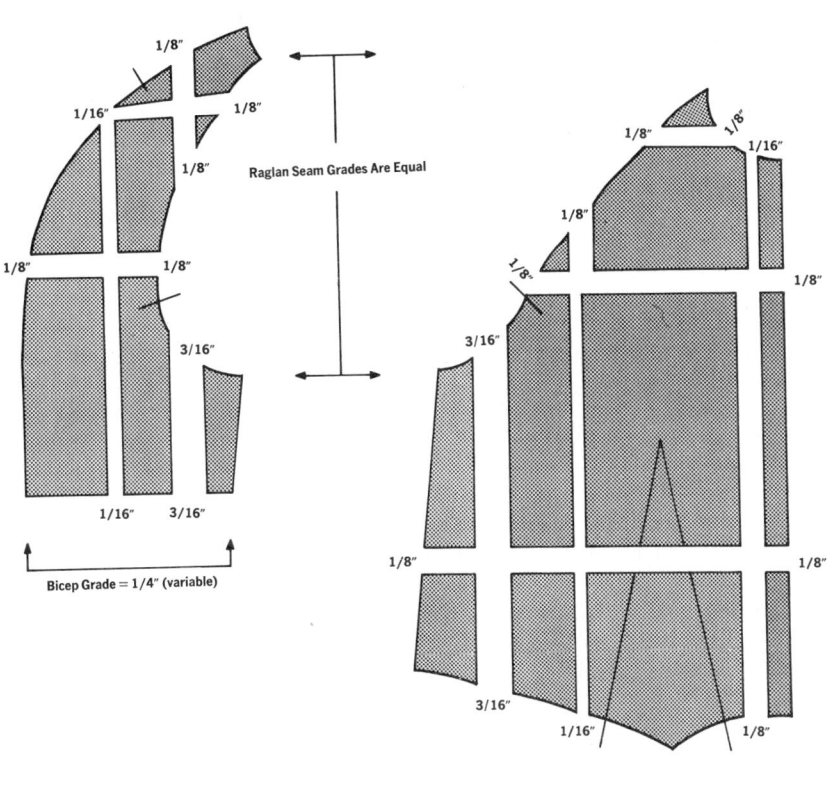

UP / OUT / IN / DOWN — These large arrows are a key to the directions in which the pattern is to be moved, as shown by the small arrows in the diagrams below.	**DIRECTIONS FOR A SET-IN RAGLAN SLEEVE** **1½ INCH GRADE—BODICE SECTION**

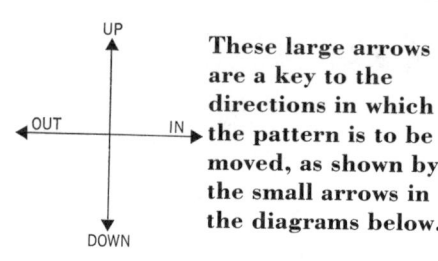

Pattern of Front Bodice

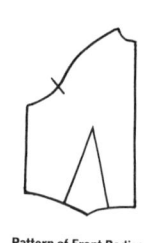

Pattern of Front Sleeve

1 CENTER FRONT

Cross Guideline / Length Guideline / Cross Control Line

---

3 NECK GRADE

Shift the bodice out 1/16″ from the length guideline. Trace the raglan seam-neckline intersection. Blend the neckline as shown on page 20.

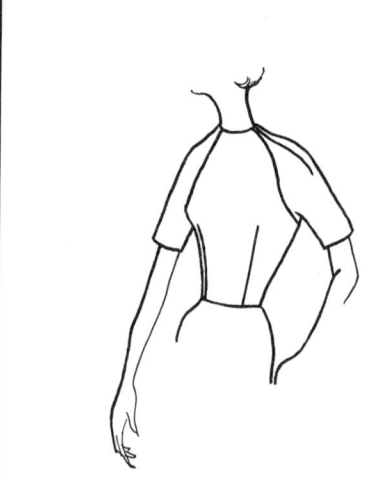

← 1/16″

4 CROSS SHOULDER GRADE*

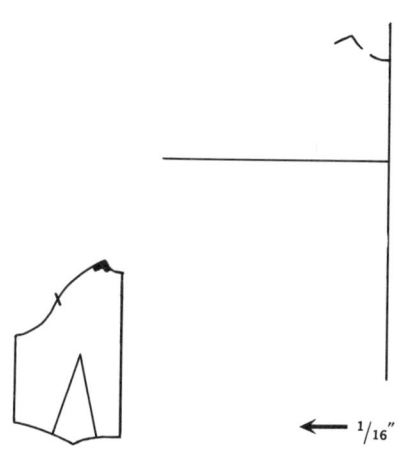

← 1/8″

then

↓ 1/16″

---

6 SIDE SEAM GRADE

Shift the bodice pattern down 1/8″ from its previous position. The cross control line of the pattern should now be a total of ¼″ below the cross guideline on the paper. Trace the side seam to and around the waistline intersection.

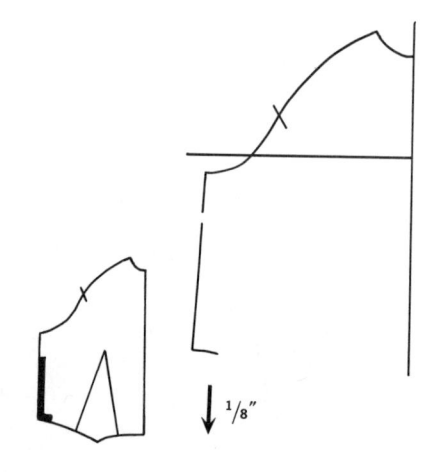

↓ 1/8″

7 APEX WIDTH GRADE

within 1/8″ →

1. In all steps except the first shift of step 4* and the second shift of step 5*, the measurements used are standard from one size to the next no matter if it be a 1", 1½", or 2" grade. Refer to the bodice grading chart for the measurements for the first shift of step 4 and the second shift of step 5 *when doing other* than the 1½" grade shown below, and follow the directions given for the front bodice.

Draw a cross control line on the pattern by squaring a line from Center Front to the side seam. On paper, draw a length guideline, and square a cross guideline from it. Place Center Front of the pattern on the length guideline, with the cross control line of the pattern on the cross guideline on the paper. Trace the neckline-Center Front intersection. (The length guideline will be the Center Front of the graded pattern).

2 SHOULDER LEVEL GRADE

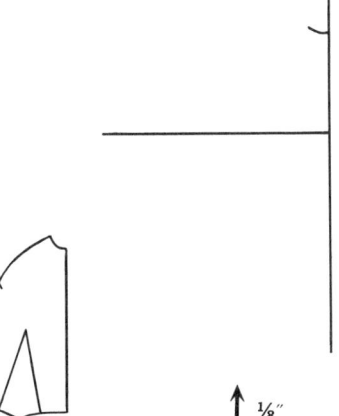

↑ ⅛"

Shift the bodice pattern up on the length guideline, until the cross control line of the pattern is ⅛" above the cross guideline on the paper. Trace nothing.

Shift the bodice pattern out ⅛" from its previous position. On a 1½" grade, the pattern should now be a total of ³⁄₁₆" from the length guideline. Trace nothing.

Then shift the pattern down ¹⁄₁₆" from its previous position. The cross control line of the pattern should now be a total of ¹⁄₁₆" above the cross guideline on the paper. Trace the middle of the raglan seam.

5 ARMHOLE AND CROSS BUST GRADE*

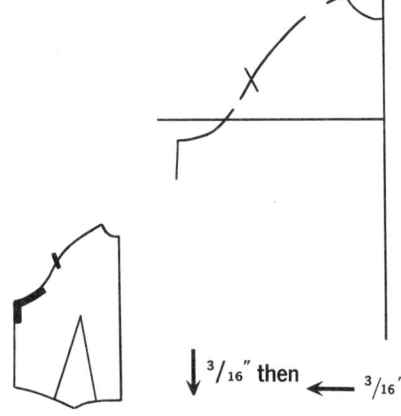

↓ ³⁄₁₆" then ← ³⁄₁₆"

Shift the bodice pattern down so that the cross control line of the pattern is ⅛" below the cross guideline on the paper. The total downward shift of the pattern from its previous position will be ³⁄₁₆". Trace nothing.

Then shift the pattern out ³⁄₁₆" from its previous position. On a 1½" grade, the pattern should now be a total of ⅜" from the length guideline. Trace the lower armhole to and around the side seam intersection. Blend the raglan seam and mark the front notch.

Shift the bodice pattern in until it is within ⅛" of the length guideline. Trace the waistline, and mark the dart notches. Blend the waistline across to the length guideline, thus establishing the new Center Front-waistline intersection.

8 DART LENGTH GRADE

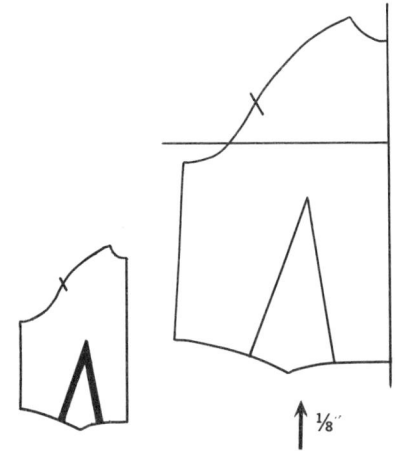

↑ ⅛"

With the bodice pattern still within ⅛" of the length guideline, shift the pattern up ⅛" from its previous position and mark the dart apex. Lift the pattern off the paper and draw in the dart lines.

The back set-in raglan bodice pattern is graded in a similar manner.

These large arrows are a key to the directions in which the pattern is to be moved, as shown by the small arrows in the diagrams below.

DIRECTIONS FOR A SET-IN RAGLAN SLEEVE
1½ INCH GRADE—SLEEVE SECTION

1 SLEEVE PREPARATION

Draw a cross control line on the pattern by ruling a line across the bicep. On paper, draw a length guideline and square a cross guideline from it. Place the cross control line of the pattern on the cross guideline of the paper, with the overarm seam of the sleeve touching the length guideline.

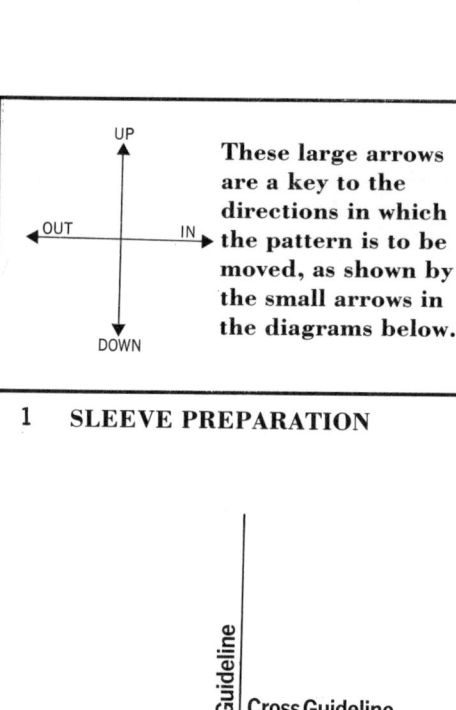

2 BICEP GRADE I*

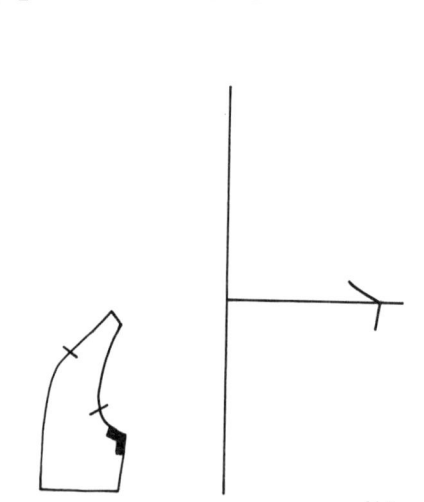

→ 1/4″

4 CAP HEIGHT GRADE

Shift the sleeve pattern up until the cross guideline of the pattern is 1/8″ above the cross guideline on the paper. Trace the lower section of the raglan seam.

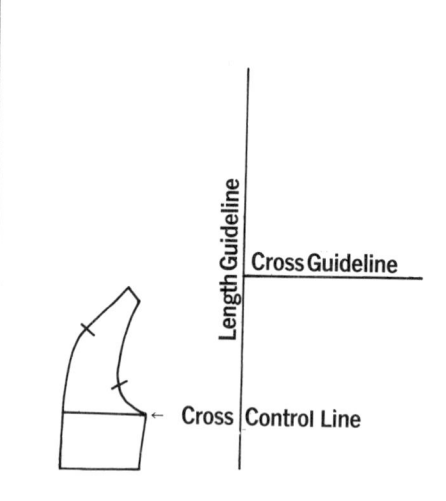

↑ 1/8″

5 CROSS SHOULDER GRADE*

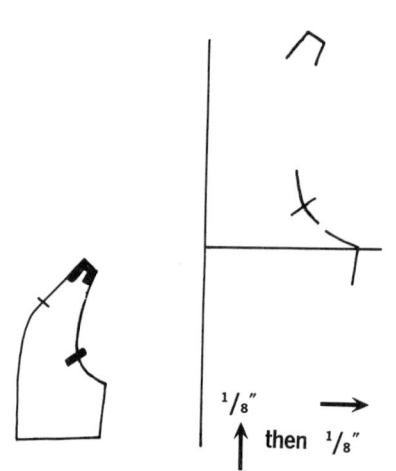

1/8″ →
↑ then 1/8″

7 OVERARM SEAM GRADE I

Shift the sleeve pattern out 1/16″ from its previous position. The overarm seam of the pattern will now be touching the length guideline on the paper. Blend the overarm seam from the bicep level to the shoulder notch.

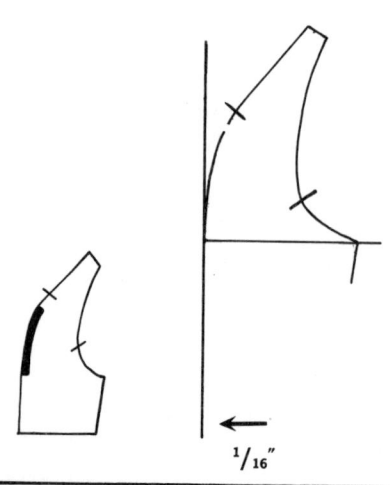

← 1/16″

8 OVERARM SEAM GRADE II

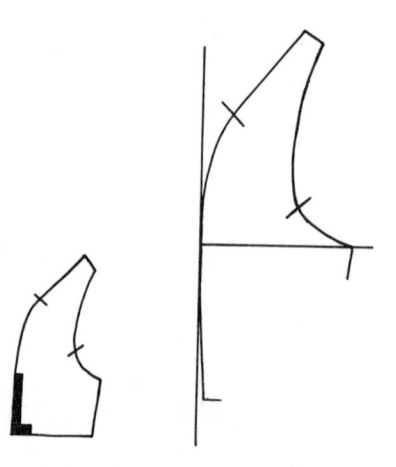

1. The measurement used in step 2* is always one-half of the total bicep grade found in column 1 on the sleeve grading chart.

2. The measurement of step 9* always equals the measurement of step 2.

3. The measurement of the second shift of step 5* and first shift of step 6* should always be equal to the cross shoulder grade used in step 4 of the raglan bodice.

4. The smaller diagram in the lower corner of each box depicts the pattern being graded, with the section to be traced in that step marked in heavy lines.

Shift the sleeve pattern in ¼″ from the length guideline. Trace the front bicep corner.

3 BICEP GRADE II

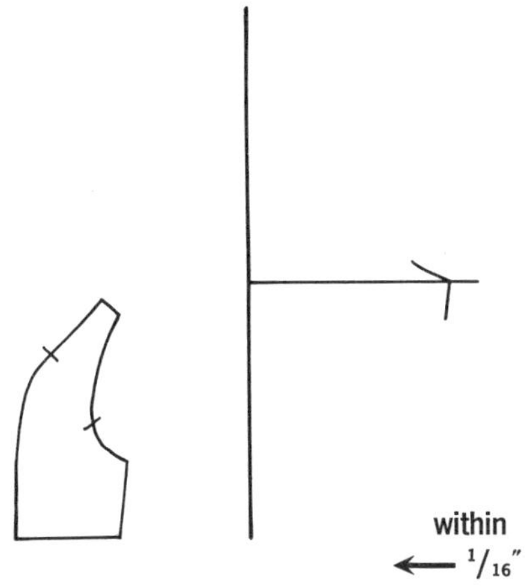

Shift the sleeve pattern out until the overarm seam of the pattern is within 1/16″ of the length guideline on the paper. Trace nothing.

Shift the pattern up ⅛″ from its previous position. The cross control line of the pattern should now be a total of ¼″ above the cross guideline on the paper. Trace nothing.

Then shift the pattern in ⅛″ from its previous position. On a 1½″ grade, the overarm seam of the pattern should now be a total of 3/16″ from the length control line on the paper. Trace the rest of the raglan seam to and around the neck and continue tracing around the neck-shoulder intersection for a short distance.

6 SHOULDER GRADE*

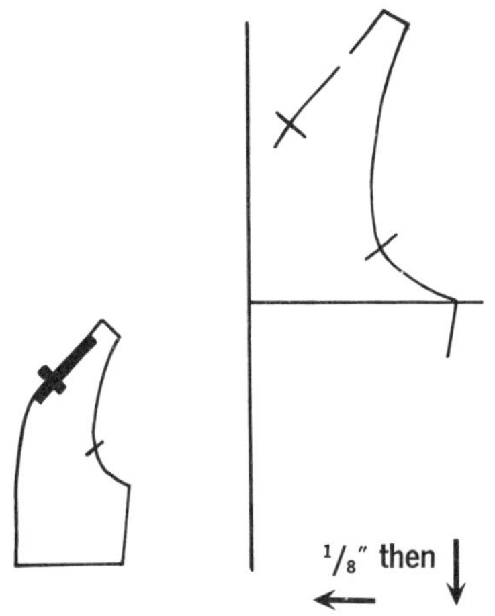

Shift the sleeve pattern out ⅛″ from its previous position, maintaining shoulder level. Trace the shoulder seam to just past the shoulder notch. Mark the notch. The overarm seam of the pattern should now be 1/16″ from the length guideline.

Then shift the pattern down until the cross control line of the pattern lies on top of the cross guideline on the paper. Trace nothing.

A short sleeve should not be graded in length. Trace the overarm seam from the bicep level to and around the bottom of the sleeve.

9 UNDERARM GRADE*

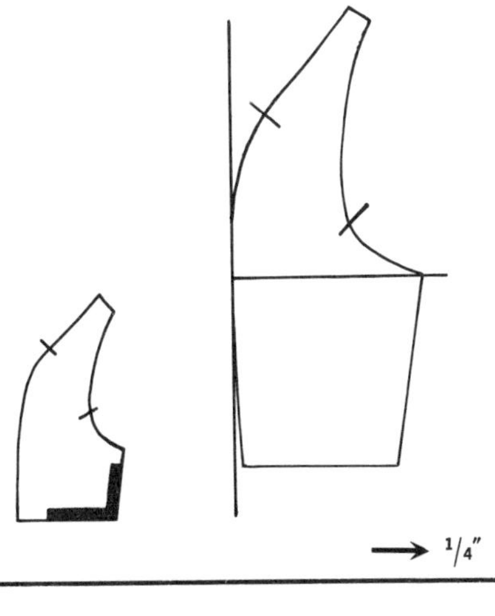

Shift the sleeve pattern out ¼″ from its previous position. The overarm seam at the cross control line should now be a total of ¼″ from the length guideline on the paper. Trace the bottom of the sleeve to and around the underarm seam up to the bicep intersection.

The back half of the set-in raglan sleeve is graded in the same manner as the front.

HOW THE BODY GROWS

PRINCESS BODICE AND SLEEVE IN ONE

1½″ Grade

1. The bodice waistline from Center Front to the princess seam is graded ⅛″, with the remainder of the waistline grade being given to the side princess section.
2. A short sleeve should not be graded in length, as shown in the diagram above and the lesson that follows.

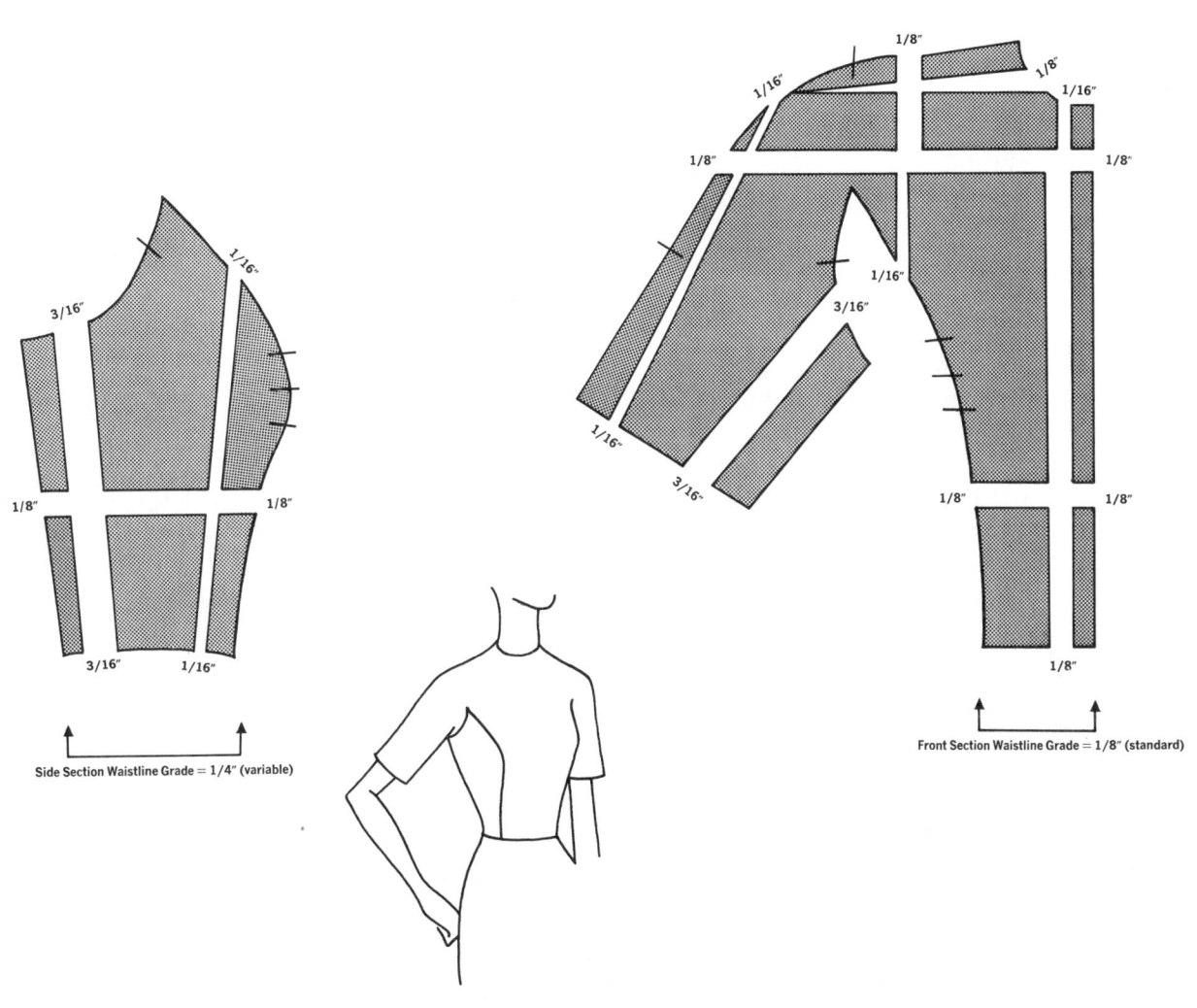

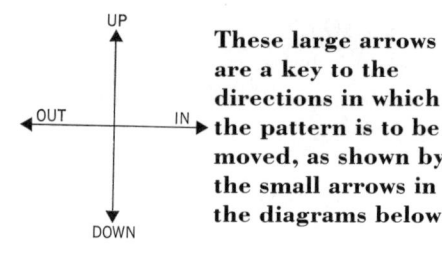

These large arrows are a key to the directions in which the pattern is to be moved, as shown by the small arrows in the diagrams below.

DIRECTIONS FOR A FRONT PRINCESS BODICE AND SLEEVE IN ONE—1½″ GRADE

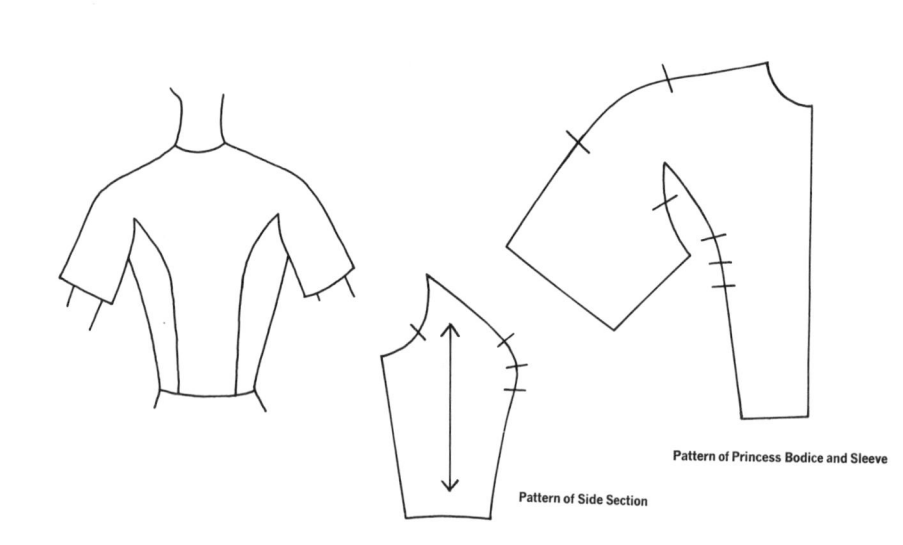

Pattern of Side Section

Pattern of Princess Bodice and Sleeve

1 CENTER FRONT

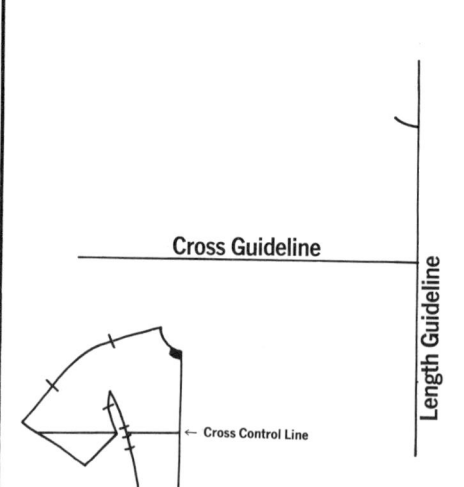

3 NECK GRADE

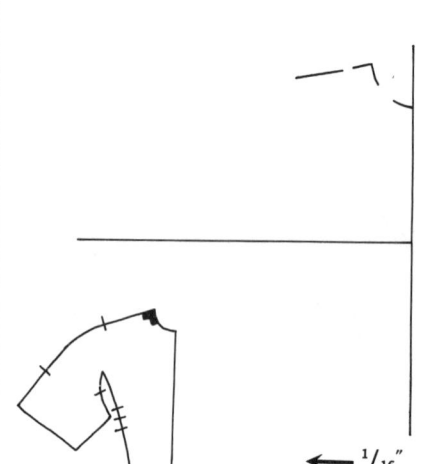

Shift the pattern out 1/16″ from the length guideline. Before tracing, adjust the pattern so that the shoulder seam is once again directly on the shoulder level traced in Step 2. This adjustment is known as "maintaining shoulder level." Now trace the shoulder—neckline intersection. Blend the neckline as shown on page 20.

4 CROSS SHOULDER GRADE*

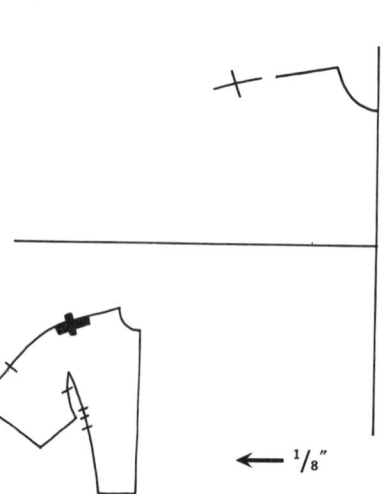

6 SHOULDER DEPTH GRADE—II

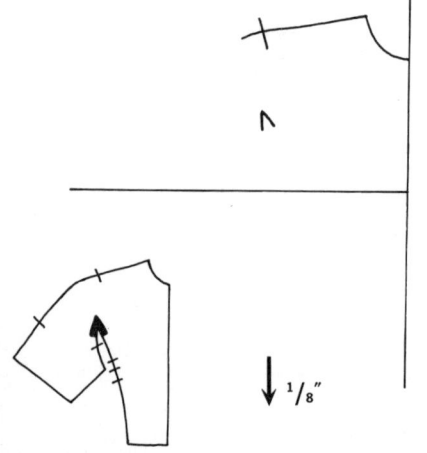

Shift the pattern down 1/8″ from its previous position. The pattern should now be a total of 1/8″ below the cross guideline on the paper. Trace the sleeve-bodice intersection for a short distance on both sides of the intersection.

7 SLEEVE WIDTH GRADE

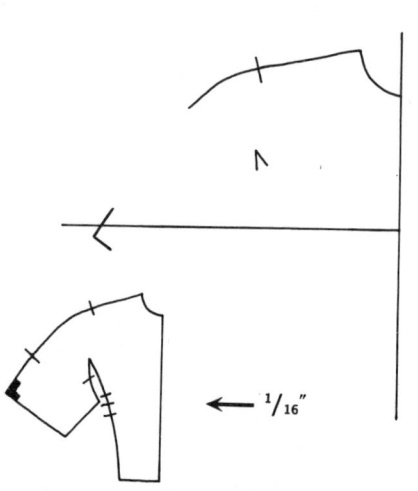

1. In all steps except 4*, the measurements used are standard from one size to the next, no matter if it be a 1", 1½", or 2" grade. Refer to the bodice grading chart for the measurement for step 4 *when doing other* than the 1½" grade shown below, and follow the directions given for the Front Bodice.

2. The smaller diagram in the lower corner of each box depicts the pattern being graded, with the section to be traced in that step marked in heavy lines.

Draw a cross control line on the pattern by squaring a line from Center Front through the sleeve bicep corner. On paper, draw a length guideline, and square a cross guideline from it. Place Center Front of the pattern on the length guideline, with the cross control line of the pattern on the cross guideline. Trace the neckline—Center Front intersection. (The length guideline will be the Center Front of the graded pattern).

2 SHOULDER LEVEL GRADE

Shift the pattern up on the length guideline, until the cross control line of the pattern is ⅛" above the cross guideline on the paper. Trace the middle of the shoulder seam in order to establish the shoulder level.

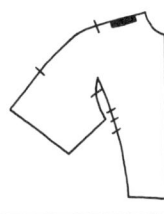

Shift the pattern out ⅛" from its previous position, maintaining shoulder level. On a 1½" grade, the pattern should now be a total of 3/16" from the length guideline. Trace the remainder of the shoulder to just past the shoulder notch. Mark the shoulder notch. (See note, step 4, Front Bodice).

5 SHOULDER DEPTH GRADE—I

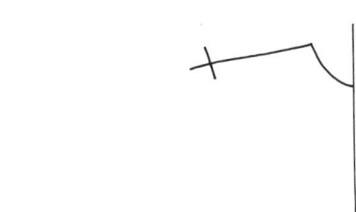

approx. ↓ 1/16"

Shift the pattern down so that the cross control line of the pattern lies on top of the cross guideline on the paper. The total shift of the pattern from its previous position will be approximately 1/16". Trace nothing.

Shift the pattern out 1/16" from its previous position. On a 1½" grade, the length control line of the pattern should now be a total of ¼" from the length guideline on the paper. Mark the overarm seam—bottom of sleeve intersection.

8 OVERARM BLEND

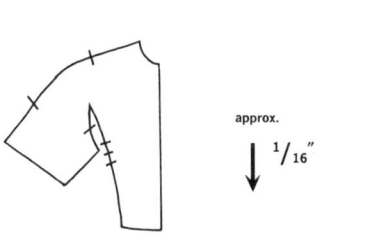

Blend the overarm seam from the bottom of the sleeve to the shoulder notch. Mark the bicep crossmark.

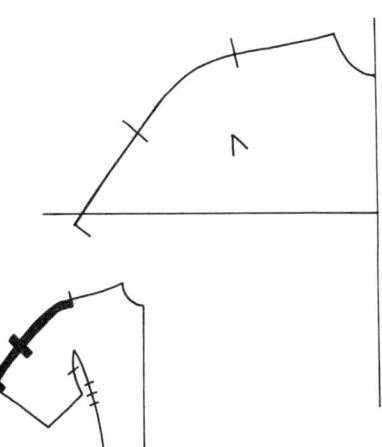

117

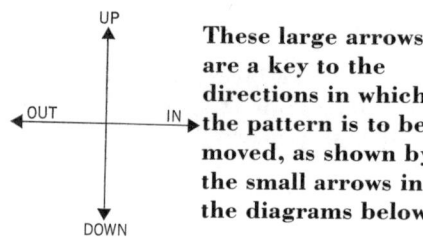

These large arrows are a key to the directions in which the pattern is to be moved, as shown by the small arrows in the diagrams below.

DIRECTIONS FOR A FRONT PRINCESS BODICE AND SLEEVE IN ONE—1½" GRADE CONTINUED

9 BOTTOM OF SLEEVE GRADE

On a 1½" grade, shift the pattern in ¼" until the Center Front of the pattern lies on top of the length guideline on the paper. For a 1" grade shift in ³⁄₁₆". For a 2" grade shift in ⁵⁄₁₆". Trace the bottom of the sleeve to and around the underarm seam intersection.

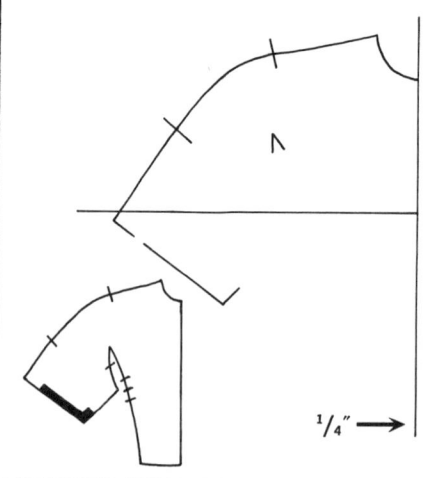

10 UNDERARM GRADE

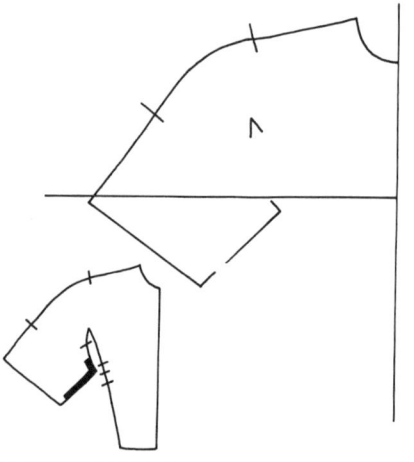

12 UPPER BODICE GRADE

Shift the pattern in until it is within ⅛" of the length guideline. Trace the princess seam between the apex notches. Mark the notches, and blend the upper princess seam. On a 1" grade, this step is omitted, as the pattern will already be within ⅛" of the length guideline.

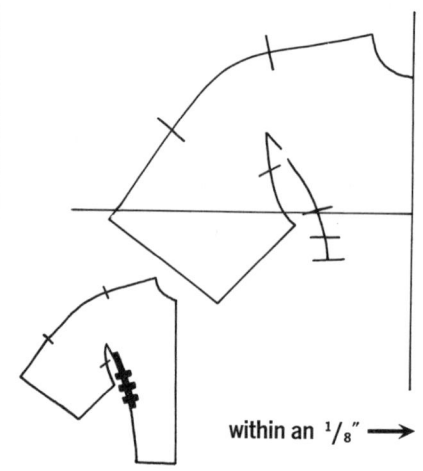

13 LOWER BODICE AND WAISTLINE GRADE

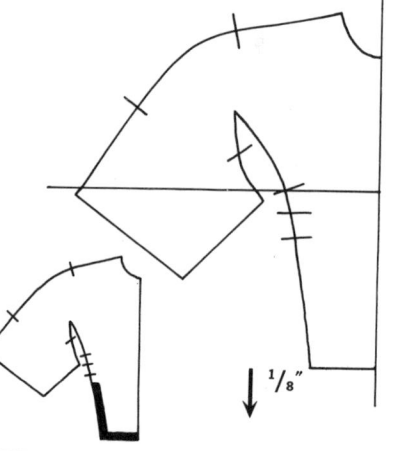

15 CROSS BUST GRADE*

Shift the pattern out ³⁄₁₆" from its previous position. For a 1½" grade, the apex of the pattern should now be a total of ¼" from the length guideline on the paper. Trace the middle of the armhole, and the side seam intersection. Blend the armhole and mark the armhole notch.

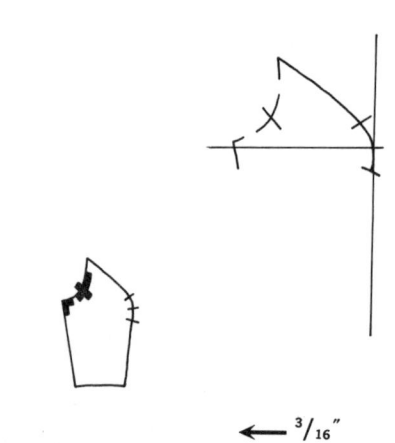

16 SIDE SEAM GRADE

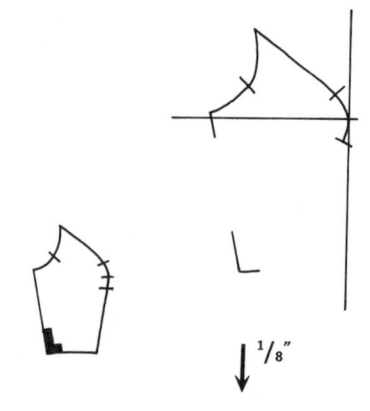

1. In all steps except Step 15*, the measurement used is standard from one size to the next, no matter if it be a 1", 1½", or 2" grade. Refer to the bodice grading chart for the measurement for step 15 *when doing other* than the 1½" grade shown below, and divide the measurement in column 6 in half before using.

2. The smaller diagram in the lower corner of each box depicts the pattern being graded, with the section to be traced in that step marked in heavy lines.

Since a short sleeve is not graded in length, trace the remainder of the underarm seam to and around the bicep intersection.

11 SLEEVE CAP GRADE

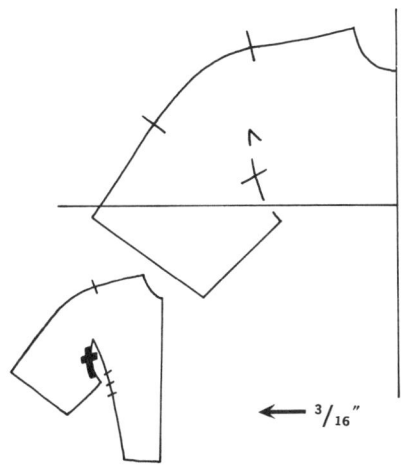

On a 1½" grade, shift the pattern out ³⁄₁₆" from the length guideline. For a 1" grade shift out ⅛". For a 2" grade shift out ¼". The sleeve—bodice intersection of the pattern should now be on top of the previously marked intersection on the paper. Trace the middle portion of the sleeve cap. Blend the sleeve cap and mark the cap notch.

Shift the pattern down ⅛" from its previous position. The cross control line of the pattern should now be ¼" below the cross guideline on the paper. Trace the remainder of the princess seam and the waistline to Center Front. Blend the waistline across to the length guideline, thus establishing the new Center Front—waistline intersection.

14 SIDE SECTION

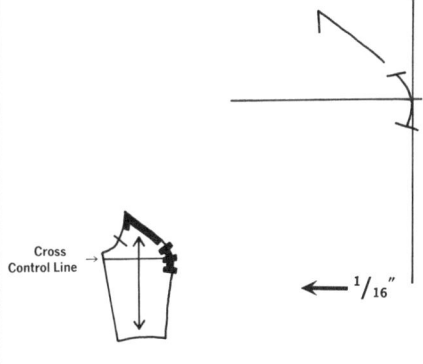

Draw a cross control line on the pattern through the apex and at a right angle to the grain line. On paper, draw a length guideline and square a cross guideline from it. Place the cross control line of the pattern on the cross guideline on the paper, with the apex at the intersection of the length and cross guidelines. Trace the princess seam between the apex notches, and mark the notches.

Then shift the pattern out ¹⁄₁₆" from the length guideline. For a 1" grade, omit this outward shift. For a 2" grade, shift out ⅛". Trace the upper princess seam to and around the underarm intersection. Blend the princess seam to the apex notch.

Shift the pattern down until the cross control line of the pattern is ⅛" below the cross guideline on the paper. Trace the side seam-waistline intersection, and blend the side seam.

17 WAISTLINE GRADE

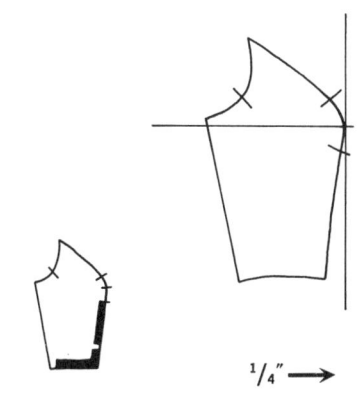

Shift the pattern in until the apex of the pattern touches the length guideline. For a 1½" grade, the total shift in of the pattern from its previous position will be ¼". Trace the waistline to and around the lower princess seam intersection. Blend the princess seam to the apex. The back princess bodice and sleeve in one is graded in a similar manner.

HOW THE BODY GROWS

DROPPED SHOULDER

1½" Grade

1. Since the cap of the sleeve has become part of the bodice, the grade usually given in the cap of the sleeve is here given to the dropped shoulder section of the bodice.

2. When grading a design, it is always useful to return to and study the grading of the original slopers from which the design was created, in order to help you visualize how the design should be graded. The dropped shoulder lesson is a good example on which to practice this rule.

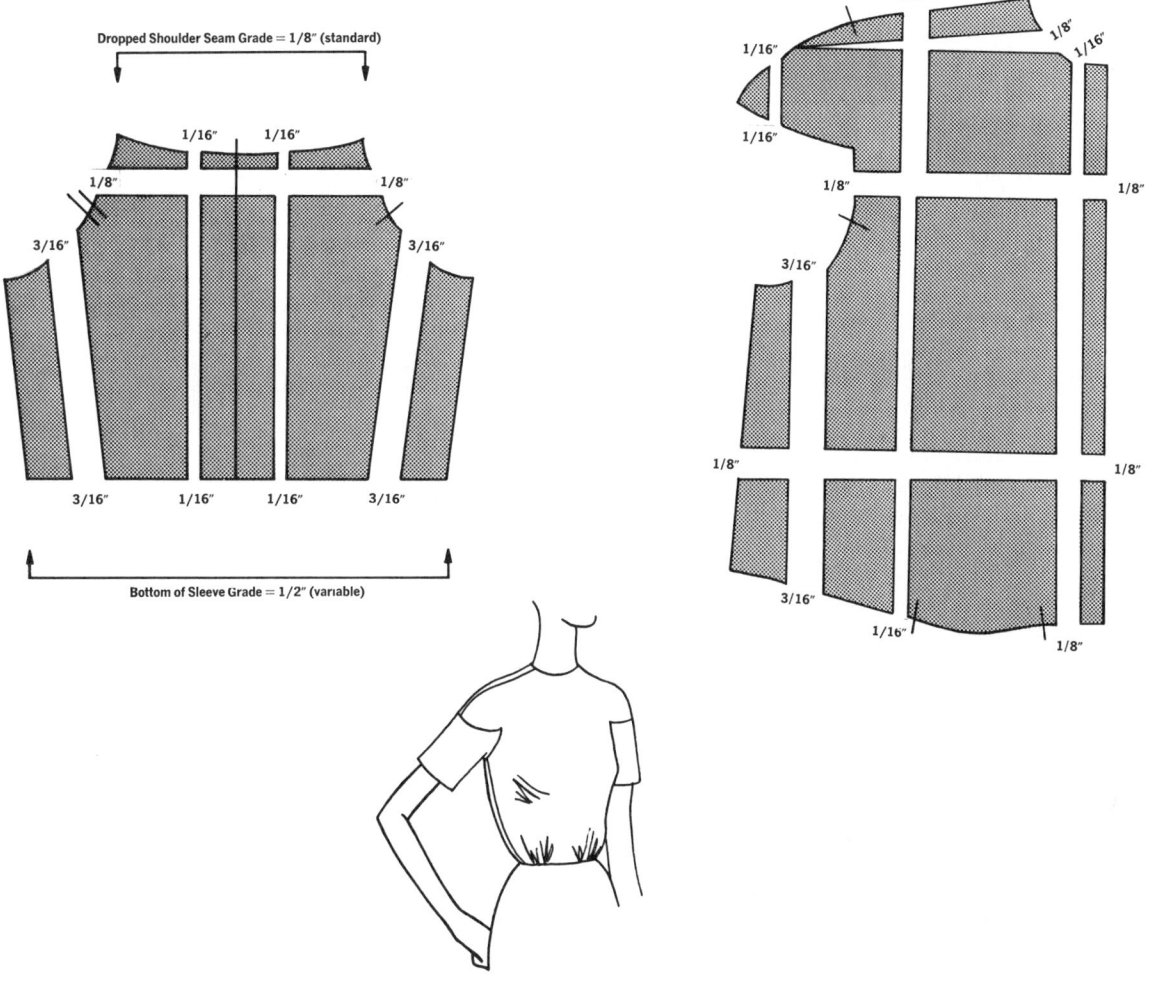

121

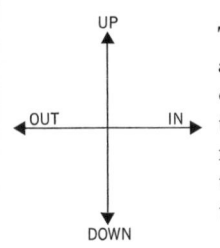

These large arrows are a key to the directions in which the pattern is to be moved, as shown by the small arrows in the diagrams below.

DIRECTIONS FOR A DROPPED SHOULDER FRONT BODICE 1½″ GRADE

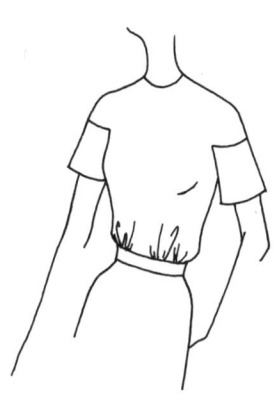

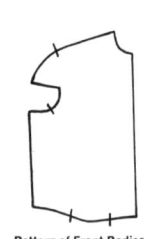

Pattern of Front Bodice

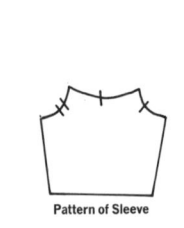

Pattern of Sleeve

1 CENTER FRONT AND SHOULDER LEVEL GRADE

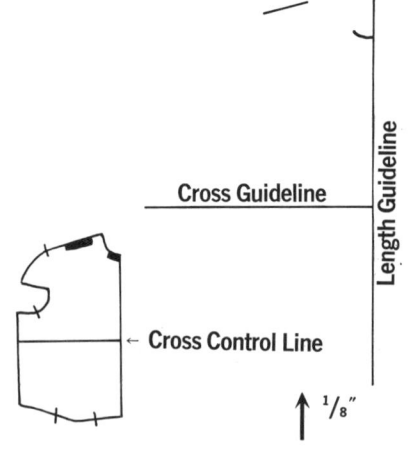

3 CROSS SHOULDER GRADE* Shift the pattern out ⅛″ from its previous position, maintaining shoulder level. The pattern should now be a total of ³⁄₁₆″ from the length guideline. Trace the shoulder seam to just past the shoulder notch. Mark the shoulder notch. (See note, step 4, Front Bodice).

← ⅛″

4 OVERARM GRADE

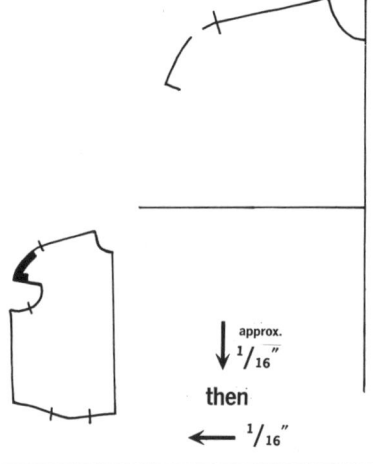

↓ approx. ¹⁄₁₆″
then
← ¹⁄₁₆″

6 CROSS BUST GRADE* Shift the pattern down so that the cross control line of the pattern is ⅛″ below the cross guideline on the paper. Trace nothing.

Then shift the pattern out ³⁄₁₆″ from its previous position. The pattern should now be a total of ⅜″ from the length guideline. Trace the lower armhole and side seam intersection. Blend the armhole as shown on page 21. Mark the armhole notch.

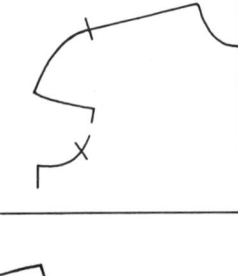

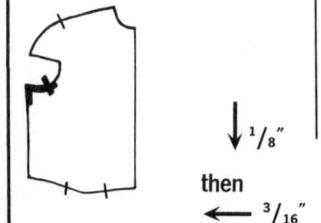

↓ ⅛″
then
← ³⁄₁₆″

7 SIDE SEAM GRADE

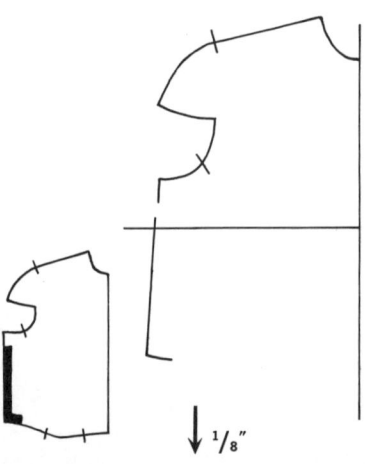

↓ ⅛″

1. In all steps except 3* and the second shift of step 6*, the measurements used are standard from one size to the next, no matter if it be a 1″, 1½″, or 2″ grade. Refer to the bodice grading chart for the measurements for step 3 and the second shift of step 6 *when doing other* than the 1½″ grade shown below. For step 3, subtract column 3 from column 4 before applying the measurement. For the second shift of step 6, divide the measurement in column 6 in half before using.

2. The smaller diagram in the lower corner of each box depicts the pattern being graded, with the section to be traced in that step marked in heavy lines.

Draw a cross control line on the pattern by squaring a line from Center Front to the side seam. On paper, draw a length guideline, and square a cross guideline from it. Place Center Front of the pattern on the length guideline, with the cross control line of the pattern on the cross guideline. Trace the neckline—Center Front intersection. (The length guideline will be the Center Front of the graded pattern).

Then shift the pattern up on the length guideline, until the cross control line of the pattern is ⅛″ above the cross guideline on the paper. Trace the middle of the shoulder seam in order to establish the shoulder level.

2 NECK GRADE

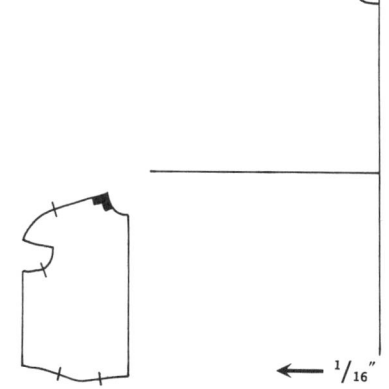

← 1/16″

Shift the pattern out 1/16″ from the length guideline. Before tracing, adjust the pattern so that the shoulder seam is once again directly on the shoulder level traced in step 1. This adjustment is known as "maintaining shoulder level." Now trace the shoulder—neckline intersection. Blend the neckline as shown on page 20.

Shift the pattern down so that the cross control line of the pattern lies on top of the cross guideline on the paper. The total shift of the pattern from its previous position will be approximately 1/16″. Trace nothing.

Then shift the pattern out 1/16″ from its previous position. The pattern should now be a total of ¼″ from the length guideline. Trace the overarm seam to and around the dropped shoulder seam intersection.

5 DROPPED SHOULDER GRADE

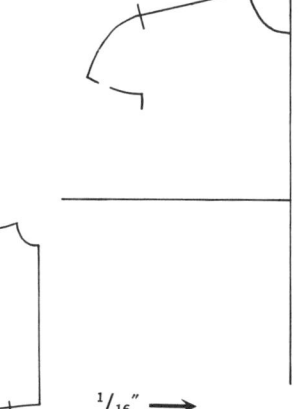

1/16″ →

Shift the pattern in 1/16″ from its previous position. The pattern should now be a total of 3/16″ from the length guideline. Trace the dropped shoulder seam to and around the armhole intersection.

Shift the pattern down ⅛″ from its previous position. The cross control line of the pattern should now be a total of ¼″ below the cross guideline on the paper. Trace the side seam to and around the waistline intersection.

8 WAISTLINE GRADE

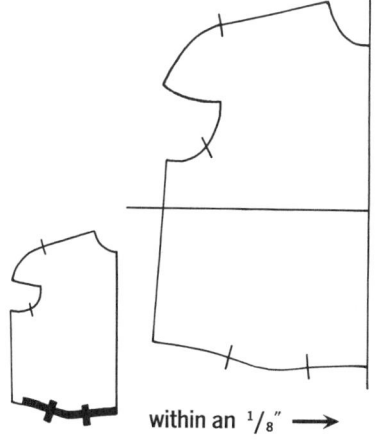

within an ⅛″ →

Shift the pattern in until it is within ⅛″ of the length guideline. Trace the waistline and the shirring notches. Blend the waistline across to the length guideline, thus establishing the new Center Front—waistline intersection. The back dropped shoulder bodice is graded in a similar manner.

These large arrows are a key to the directions in which the pattern is to be moved, as shown by the small arrows in the diagrams below.

DIRECTIONS FOR A DROPPED SHOULDER SLEEVE 1½″ GRADE

1 SLEEVE PREPARATION

Draw a cross control line on the pattern by ruling a line across the bicep. Draw a length control line on the pattern by ruling a line through the center of the sleeve. On paper, draw a length guideline and square a cross guideline from it.

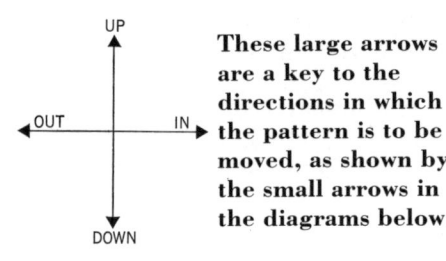

2 FRONT BICEP GRADE*

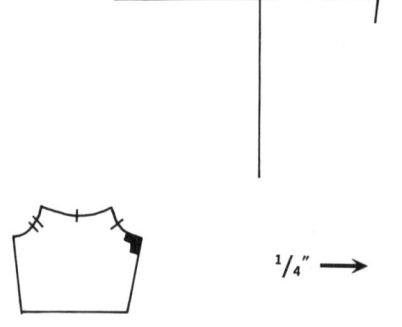

4 FRONT DROPPED SHOULDER SEAM GRADE

Shift the pattern out ¹⁄₁₆″ until the length control line of the pattern lies on top of the length guideline on the paper. Trace the front dropped shoulder seam, and mark the center sleeve notch.

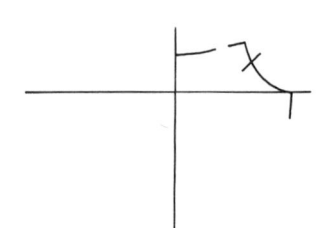

5 BACK DROPPED SHOULDER SEAM GRADE

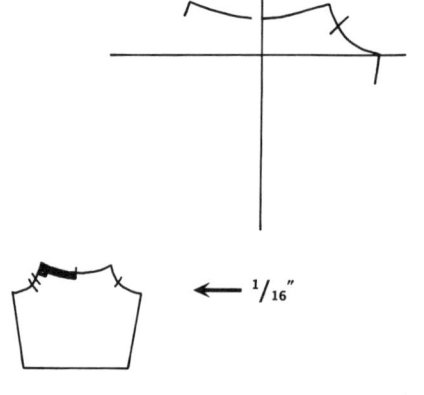

7 BACK BICEP GRADE*

Shift the pattern out ³⁄₁₆″ from its previous position. On a 1½″ grade, the length control line of the pattern should now be a total of ¼″ out from the length guideline on the paper. Trace the back bicep corner. Blend the sleeve cap and mark the back cap notches.

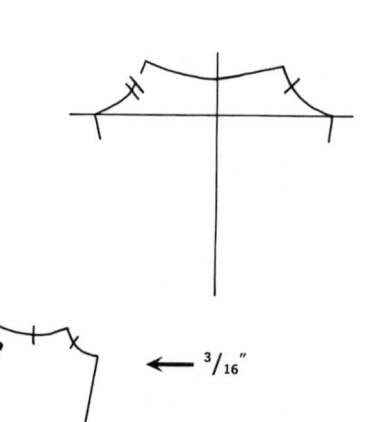

8 BACK UNDERARM SEAM GRADE

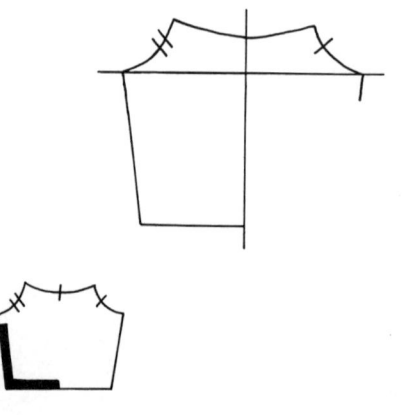

1. In all steps except 2* and 7*, the measurements used are standard from one size to the next, no matter if it be a 1", 1½", or 2" grade. Refer to the sleeve grading chart for the measurements for steps 2 and 7 *when doing other than the 1½" grade shown below.* For step 2, divide the measurement in column 6 in half before using. For step 7, divide the measurement in half, and subtract ¹/₁₆" before using.

2. The smaller diagram in the lower corner of each box depicts the pattern being graded, with the section to be traced in that step marked in heavy lines.

Place the pattern with its control lines on top of the corresponding guidelines on the paper. Shift the pattern in ¼" from the length guideline. Trace the front bicep corner.

3 FRONT CAP GRADE

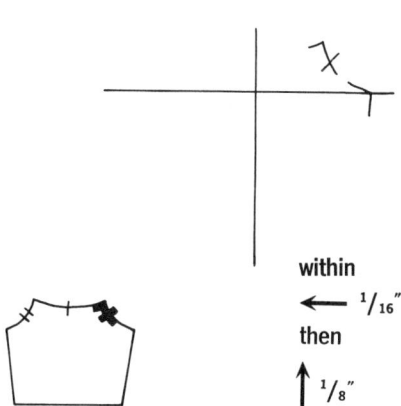

Shift the pattern out until the length control line of the pattern is within ¹/₁₆" of the length guideline on the paper. Trace nothing.

Then shift the pattern up until the cross control line of the pattern is ⅛" above the cross guideline on the paper. Trace the front cap seam to and around the dropped shoulder seam intersection. Blend the front cap and mark the front cap notch.

Shift the pattern out ¹/₁₆" from the length guideline. Trace the back dropped shoulder seam to and around the back cap intersection.

6 BACK CAP GRADE

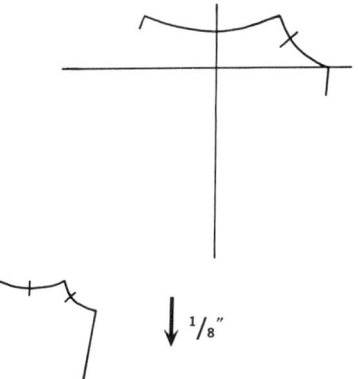

Shift the pattern down ⅛" until the cross control line of the pattern lies on top of the cross guideline on the paper. Trace nothing.

A short sleeve does not get graded in length. Trace the back underarm seam to and around the bottom of the sleeve to the length control line.

9 FRONT UNDERARM SEAM GRADE

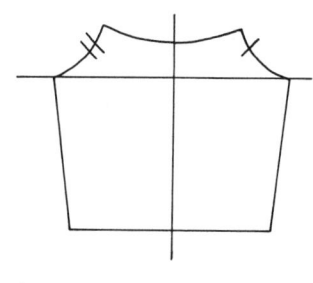

Shift the pattern in ½" until the length control line of the pattern is ¼" out from the length guideline on the paper. Trace the remainder of the bottom of the sleeve to and around the front underarm seam and up to the front bicep intersection.

NOTE: The inward shift of this step is always equal to the total bicep grade of the sleeve.

For a 1" grade, shift in ⅜".
For a 2" grade, shift in ⅝".

ADDITIONAL DESIGNS FOR GRADING

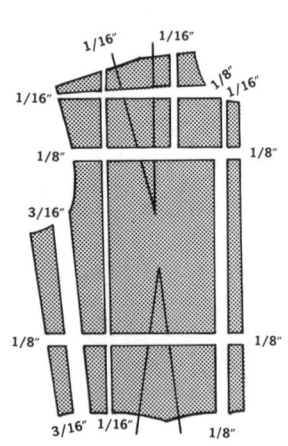

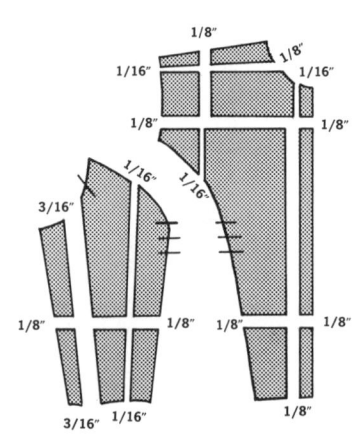

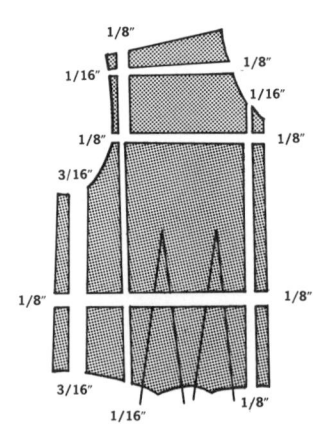

FRONT BODICE WITH SHOULDER DART

A front bodice with a shoulder and a waistline dart follows the same procedure as the one dart bodice, except the total shoulder grade is divided equally on either side of the shoulder dart.

ARMHOLE PRINCESS LINE BODICE

In this split diagram, the bodice sections have been aligned as they would be joined, showing how the growth is carried through from the center panel to the side section. The grading procedure is similar to the princess bodice and sleeve in one.

BODICE WITH MULTIPLE DARTS OR PLEATS

A front bodice with multiple darts or pleats grades the same as a bodice with a single dart. The waistline from Center Front to the first dart is graded an 1/8″, with the remainder of the waistline grade being given on the other side of the multiple darts or pleats.

A V-plunge neckline is not graded in length more than the standard neckline increase.

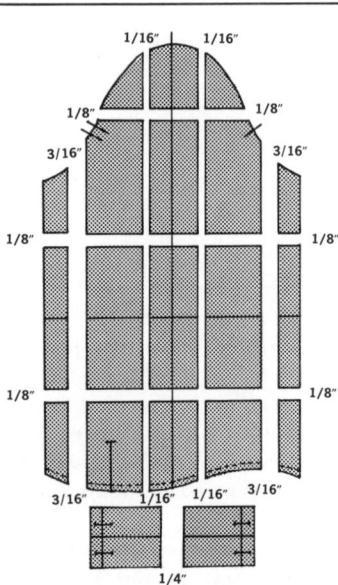

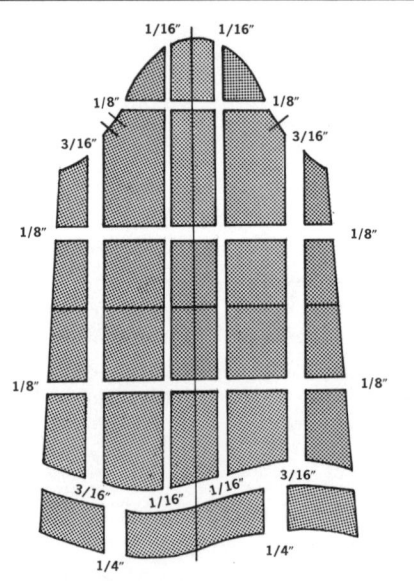

SHIRTWAIST SLEEVE WITH FRENCH CUFF

The shirtwaist sleeve follows the same procedure as a set-in fitted sleeve except that the wrist is graded the same as the bicep. The french cuff grades 1/4″, the same grade as the wrist of a fitted set-in sleeve. The width of the cuff remains the same for all sizes.

BISHOP SLEEVE WITH CUFF

This sleeve is graded the same as the shirtwaist sleeve. All fitted cuffs attached to a gathered sleeve grade in length an amount equal to the fitted wrist grade, without changing the original width of the cuff.

BELL SLEEVE WITH FACING

As in the preceding split diagrams, the grade of the bell sleeve can be compared with that of the shirtwaist sleeve.

The grade of all facings is the same amount as the edge to which it is attached, with its width remaining the same.

Now that you have had an opportunity to practice the grading techniques in this book by the step-by-step method, the split diagrams below should serve as a guide for additional grading on your own. The diagrams illustrate the 1½″ grade.

ADDITIONAL DESIGNS FOR GRADING

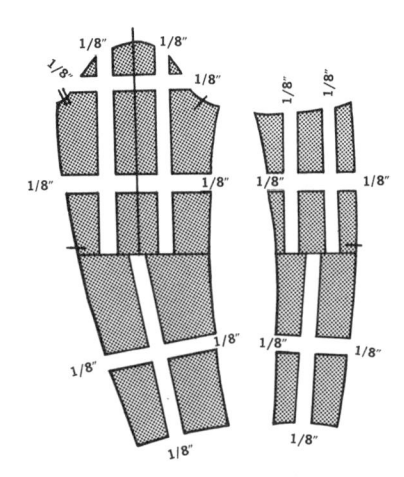

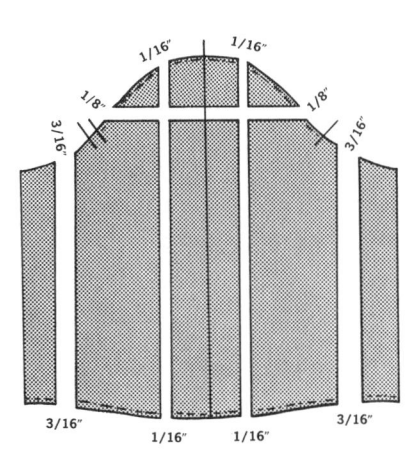

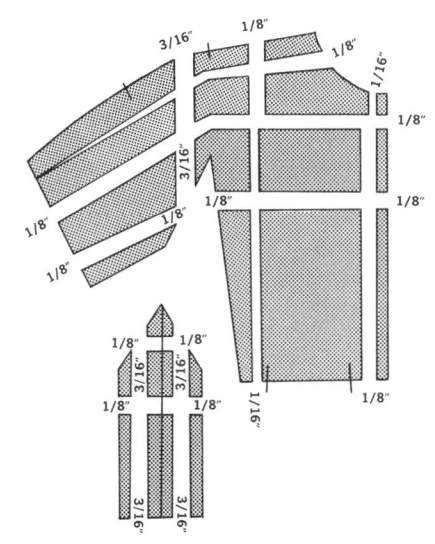

THE TWO-PIECE SLEEVE
The two-piece sleeve as a whole grades the same amount as the one-piece sleeve. The grading measurements are distributed as shown between the overarm and underarm sections. If a placket closing were used, it would remain the same for all sizes.

THE PUFF SLEEVE
A short puff sleeve need not be graded in length. The grade at the bottom of the sleeve should equal the grade of the bicep. A short sleeve, whether puffed or belled, is graded in the same manner as the set-in sleeve.

THE HOUSE GUSSET
This variation on the gusset is graded similarly to the diamond gusset, and that lesson may be used as a guide for the grading of the house gusset, as this variation is called. The diagram shows both the front and back sections of the gusset joined at the sideseam.

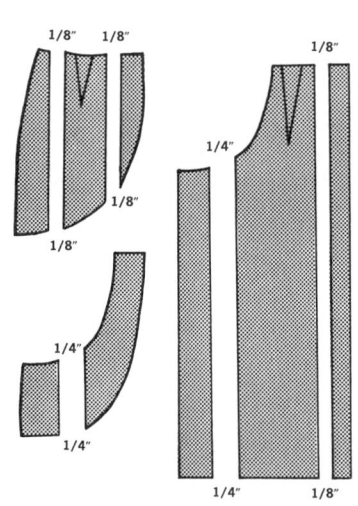

SKIRT POCKET
The split diagram shows the skirt, the pocket, and the pocket facing, with the skirt section held in length. While pockets such as the one illustrated are affected by the grade of the skirt as a whole, separate details such as a patch pocket do not receive any grade.

NOTCH COLLAR
The split diagram shows a notched collar with a double breasted closing. Note that the grade of the double breasted extension is given to the collar along the corresponding seam, in addition to the standard collar neck grade. Note also that the center front grade is given entirely below the lapel, so as not to affect the neckline grade.

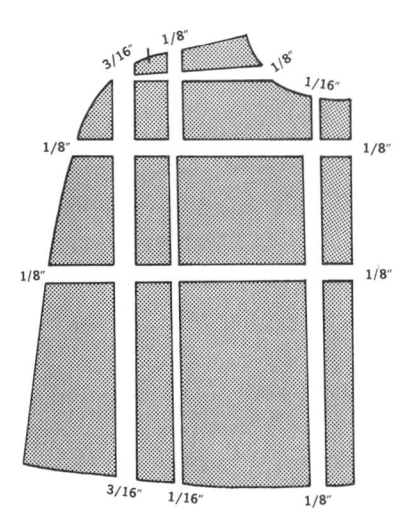

CAPES
The grade of the cape is based upon the grade of the bodice, as a study of the diagram will show. Other than the standard center front grade, the cape may be held in length, as illustrated. Details, such as slits for the arms, are not graded.

INDEX

Accuracy of grading	13, 24
Apex width grade	14-15
Armhole depth grade	14-15, 25
Assembling pattern pieces	24
Bell bottom slacks	45-47
Bell sleeve	126
Bicep grade	16-17
description of	4-5
Bicep to wrist grade	16-17
Bishop sleeve	126
Blending	19-21
armhole	21
collars	20
corners	20
crotch seam	20
curves	20-21
neckline	20
seam with notch	21
sleeve cap	20
sleeve underarm	19
shoulder seam	19
straight lines	19
Blouse, tuck-in	41-43
Bodice	
basic back	
neck dart	33-35
shoulder dart	29-31
basic front	
waist dart	25-27
princess, shoulder	63-65
shirring	55, 59, 79, 87
Body landmarks	4-6
circumference	4
length	4
width	4
Body measurements,	
definition of	2-3
Body types,	
analysis of	8-10
Bust dart	41-43
Bust grade	5, 11
Junior	14
Misses	15
Buttons and buttonholes	79
Cap height grade	16-17, 49
Cape	127
Center Back,	
neck to waistline grade	4
Center Front,	
neck to waistline grade	4
Circle skirt	71-73
Circumference grade	5
Collars	37-39
buster brown	37-39
convertible	37-39
eton	37-39
mandarin	37-39
notched collar	127
peter pan	37-39
sailor	37-39
shawl	77-79
Collar neckline	37
Collar width	37
Cross shoulder grade	
description of	4, 6, 25
grade for Juniors	14
grade for Misses	15
Cuffs	126
Darts	
back neck	33-35
back shoulder	29-31
elbow	49-51
length grade	14-15, 41
multiple	45-47, 37-39, 126
pick-up	41
side bust dart	41-43
waistline	25-27
Definitions	2-3
Downgrade	12
Dirndl skirt	74
Double breasted closing	77, 127
Dropped shoulder	121-125
Eight gore skirt	74
Elbow dart length grade	16-17, 49
Elbow grade	
description of	4
grade of	16-17, 49
grade for kimono sleeves	83, 87
Facings	126
Figure types	8-10
Fit	1-2
Fitted set-in sleeve	49-51
Flares	75
Four gore skirt	74
French cuff	126
Girth	3
Gored skirts	
eight gore	74
four gore	74
gathered front six gore	75
six gore pleated skirt	75
trumpet	75

Grading	
definition of	2
importance of	1
Grading charts	
use of	11-12
bodice, Junior	14
bodice, Misses	15
downgrade	12
skipping sizes	13
sleeve, Junior	16
sleeve, Misses	17
skirt, Junior	16
skirt, Misses	17
Grading machine	18, 65
Grade types	
circumference	5
length	6
width	6
Gussets	
house	127
two piece	101, 106-107
Half-size body type	8-10
Height of figure types	8
Hem flare	
circle skirts	71
basic skirts	37
Hip grade	
description of	4, 6, 37
Hip to hemline grade	
description of	4, 6, 37
Hip yoke	75
House Gusset	127
Junior body type	8-10
Junior grading charts	
bodice	14
skirt	16
sleeve	16
Junior petite body type	8-10
Kimono sleeve	83-87
with a gusset	101-107
with a raglan seam	89-93
with a square armhole	95-99
Length grade	6
Machines, grading	18
Mandarin collar	37-39
Master pattern	2
Midriff, bodice	59-61
Misses body type	8-9
Misses grading charts	
bodice	15
skirt	17
sleeve	17
Multiple darts	
bodice	126
skirt	37-39
slacks	45-47
torso	41-43
Neck dart, back bodice	33-35
Neck grade	
description of	4, 6, 25
grade of	14-15
V-neckline	126
Notched collar	127
One-piece dress	41-43
Overblouse	41-43
Overarm seam	4
Peg top skirts	74
Peter pan collar	
flat and rolled	37-39
Pleats	
bodice	126
skirts	75
Pockets	
in skirt	127
patch	127
Princess bodice	
and sleeve in one	115-119
Princess line bodice	
armhole	126
shoulder	63-65
Production pattern	2
Puffed sleeve	127
Raglan sleeve	
kimono	89-93
set-in	109-113
Sailor collar	37-39
Separate set-in collars	
buster brown	37-39
convertible	37-39
eton	37-39
mandarin	37-39
peter pan, flat or rolled	37-39
sailor	37-39

Set-in raglan sleeve	109-113
Set-in sleeve	49-51
bell	126
bishop	126
puffed	127
shirtwaist	126
short fitted	115
straight	49
Shawl collar	77-79
Shirring	
bodice	55, 59, 79, 87
skirt	74-75
Shirtwaist sleeve	126
Shorts	45-47
Shoulder dart	
back bodice	29-31
front bodice	126
Shoulder level grade	
description of	25-26
grade of	14-15
Side bust dart	41-43
Side seam to hemline	4
Side seam to waist	
description of	4, 6
grade of	14-15
Six gore skirt	67-69
Size range	
definition of	3
description of	8-10
Skipping sizes	13
Skirts	
circular	71-73
dirndl	74
gored	74-75
hip yoke	75
peg top	74
pleats	75
skirt with pockets	127
trumpet	75
Skirt length	37, 74
Slacks	
bell bottom	45-47
straight leg	45-47
tapered leg	45-47
Sleeves	
bell	126
bishop	126
dropped shoulder	
sleeve	121-125
gusset sleeve	101-107
puffed sleeve	127
princess bodice and	
sleeve in one	115-119
kimono raglan	
sleeve	89-93
kimono sleeve	83-87
set-in fitted sleeve	49-51
set-in raglan sleeve	109-113
shirtwaist	126
square armhole	95-96
two-piece sleeve	127
Split diagrams	
back bodice with	
neck dart	33
back bodice with	
shoulder dart	29
bell sleeve	126
bishop sleeve	126
bodice and midriff	59
bodice and yoke	55
cape	127
circle skirt	71
collars	37
cuffs	126
dropped shoulder	
sleeve	121
facings	126
front bodice with	
multiple darts	126
pleats	126
shoulder dart	126
waist dart	25
gusset sleeve	101
house gusset	127
kimono raglan sleeve	89
kimono sleeve	83
notched collar	127
princess bodice	
armhole princess	126
shoulder princess	63
princess bodice and	
sleeve in one	115
puffed sleeve	127
set-in fitted sleeve	49
set-in raglan sleeve	109
shawl collar	77
shirtwaist sleeve	126
six gore skirt	67
skirt	37
skirt with pockets	127
slacks	45
square armhole	95
torso	41
two piece gusset	101
two piece sleeve	127
Standard measurements	11
Straight leg slacks	45-47
Style lines	121
Square armhole	95-99
Tapered leg slacks	45-47
Two-pieced gusset	101
Two-pieced sleeve	127

Uneven grade	7
Variable measurements	11
V-neckline	126
Vocabulary	2-3
Waistline grade	
description of	4-5
grade for Juniors	16
grade for Misses	17
with gores	67, 74
with multiple darts	
bodice	126
skirt	37
with multiple pleats	126

Waistline to hip grade	4, 41
Width grade	6
Wrist grade	
description of	4-5
grade of fitted wrist	16-17, 49, 83
grade of gathered wrist	126
grade of straight wrist	16-17, 49
Yokes	
bodice	55-57
hip	75